DEVELOPMENTAL BIOLOGY

DEVELOPMENTAL BIOLOGY

From a Cell to an Organism

RUSS HODGE

FOREWORD BY NADIA ROSENTHAL, PH.D.

This book is dedicated to the memory of my grandparents, E. J. and Mabel Evens and Irene Hodge, to my parents, Ed and Jo Hodge, and especially to my wife, Gabi, and my children—Jesper, Sharon, and Lisa—with love.

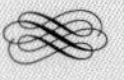

DEVELOPMENTAL BIOLOGY: From a Cell to an Organism

For information contact:

Facts On File, Inc.
An imprint of Infobase Publishing
132 West 31st Street
New York NY 10001

Library of Congress Cataloging-in-Publication Data

Hodge, Russ, 1961–
Developmental biology : from a cell to an organism / Russ Hudge ; foreword by Nadia Rosenthal.
p. cm. — (Genetics and Evolution)
Includes bibliographical references and index.
ISBN 978-0-8160-6683-4
1. Developmental biology. I. Title.
QH491.H63 2010
571.8—dc22 2008055557

Facts On File books are available at special discounts when purchased in bulk quantities for businesses, associations, institutions, or sales promotions. Please call our Special Sales Department in New York at (212) 967-8800 or (800) 322-8755.

You can find Facts On File on the World Wide Web at http://www.factsonfile.com

Text design by Kerry Casey
Illustrations by Sholto Ainslie, and Chris and Elisa Scherer
Photo research by Elizabeth H. Oakes

Printed in the United States of America

Bang Hermitage 10 9 8 7 6 5 4 3 2 1

This book is printed on acid-free paper.

I say that it touches a man that his blood is sea water and his tears are salt, that the seed of his loins is scarcely different from the same cells in a seaweed, and that of stuff like his bones coral is made. I say that the physical and biologic law lies down with him, and wakes when a child stirs in the womb, and that the sap in a tree, uprushing in the spring, and the smell of the loam, where the bacteria bestir themselves in darkness, and the path of the sun in the heaven, these are facts of first importance to his mental conclusions, and that a man who goes in no consciousness of them is a drifter and a dreamer, without a home or any contact with reality.

—*An Almanac for Moderns: A Daybook of Nature,*
Donald Culross Peattie, copyright ©1935
(renewed 1963) by Donald Culross Peattie

Contents

Foreword

Few people can remember things that happened to them before the age of two or three, yet by the time a person is born, he or she has already been on a long hard journey. Human embryonic development involves acrobatic feats of rapid growth, twisting, and even turning inside out. Although we cannot recall our own experience of embryonic development, biologists are currently piecing together this journey step by step—discovering how microscopic embryos develop into complex organisms—and are consequently on the verge of solving one of the most fascinating mysteries of modern biology. Along the way, these detectives are showing how humans share many of the steps with elephants and whales, monkeys and mice, with only subtle variations that define us as a different species.

Developmental Biology by Russ Hodge is a guided tour of humankind's embryological journey. It is a story not only of the embryo but also of the triumphs and tribulations of biologists, from the earliest history of embryology to the latest theories in this fascinating field. Chapter 1 discusses the initial moments of an organism's life cycle and follows how one cell, the fertilized egg, gives rise to millions of cells using information embedded in its genetic material. The rapid divisions of this first cell, according to a precise order and time frame, are critical to the developmental potential of the resulting embryo.

Chapter 2 describes how molecular signals passed between cells at precise times and places drive the formation of early patterns in tissues. Delicate control systems inform cells in one part of an embryo to adopt one fate and those elsewhere to adopt another. Timing is crucial: Cells in embryos use molecular clocks as well as spatial signals to tell them where they are and what they should become. If any of these programs falters, the embryo is doomed. The signals must be sufficiently robust to work perfectly every time.

Pattern formation is the dominant theme of chapter 3, in which we follow how cells in the embryo continue to divide along specific blueprints to build hearts, brains, and eyes, then halt growth when a structure has reached its proper size and shape. In chapter 4 we learn what happens when developmental programs go wrong. An important theme is the way stem cells establish tissues during embryonic development and how some of these cells persist after birth to self-renew our bodies and repair injuries. Understanding more about stem cells may help in the medical battle to repair tissue damage that occurs during diseases, but we also now know that aberrant stem cells are the source of many cancers. The same characteristics that allow a stem cell to regenerate a damaged organ might explain, for example, how a tumor may form again, even after treatment with anticancer drugs, if a single "cancer stem cell" has survived.

The final chapter of the book deals with the remarkable conservation of developmental mechanisms across evolutionary time scales. Now it is a given that related genes function in similar ways to build the same types of structures in many different species. This fact became obvious to several biologists in the 19th century who had extraordinary powers of observation but no clue as to the molecules that caused embryos to develop in certain ways. Even so, by comparing embryos of different species at different stages, they discovered that evolution worked by modifying existing structures. Today we know why: Subtle changes in genes alter signals and create slight differences during development that can have big effects on a body. For example, changes in just a few genes seem to have been responsible for a great leap in brain size in primates, leading to a magnificent human brain that can now investigate its own origins.

Although modern biologists have made impressive progress in pinpointing some of the key molecules that govern how embryos generate patterns over space and time, in many ways we are only at the beginning of the quest to solve the great questions of development. How do signals arise in the embryo to organize patterns of cell growth? In the earliest phases of life an organism's cells are nearly identical, so what genetic programs make regions different from one another? With new, sophisticated tools and techniques, developmental biologists can start to decode the programs by which molecules work together to create patterns that make every embryo the same as other members of its species, yet also a bit unique.

The story of your prehistory—those early, crucial phases of life as an embryo—is not yet complete. But, *Developmental Biology* will bring you up to date with what we know and introduce some of the questions that remain for the next generation of researchers to solve.

—Nadia Rosenthal, Ph.D.
Head of Outstation,
European Molecular Biology Laboratory,
Rome, Italy

Preface

In laboratories, clinics, and companies around the world, an amazing revolution is taking place in our understanding of life. It will dramatically change the way medicine is practiced and have other effects on nearly everyone alive today. This revolution makes the news nearly every day, but the headlines often seem mysterious and scary. Discoveries are being made at such a dizzying pace that even scientists, let alone the public, can barely keep up.

The six-volume Genetics and Evolution set aims to explain what is happening in biological research and put things into perspective for high-school students and the general public. The themes are the main fields of current research devoted to four volumes: *Evolution, The Molecules of Life, Genetic Engineering,* and *Developmental Biology*. A fifth volume is devoted to *Human Genetics,* and the sixth, *The Future of Genetics,* takes a look at how these sciences are likely to shape science and society in the future. The books aim to fill an important need by connecting the history of scientific ideas and methods to their impact on today's research. *Evolution,* for example, begins by explaining why a new theory of life was necessary in the 19th century. It goes on to show how the theory is helping create new animal models of human diseases and is shedding light on the genomes of humans, other animals, and plants.

Most of what is happening in the life sciences today can be traced back to a series of discoveries made in the mid-19th century. Evolution, cell biology, heredity, chemistry, embryology, and modern medicine were born during that era. At first these fields approached life from different points of view, using different methods. But they have steadily grown closer, and today they are all coming together in a view of life that stretches from single molecules to whole organisms, complex interactions between species, and the environment.

The meeting point of these traditions is the cell. Over the last 50 years biochemists have learned how DNA, RNA, and proteins carry out a complex dialogue with the environment to manage the cell's daily business and to build complex organisms. Medicine is also focusing on cells: Bacteria and viruses cause damage by invading cells and disrupting what is going on inside. Other diseases—such as cancer or Alzheimer's disease—arise from inherent defects in cells that we may soon learn to repair.

This is a change in orientation. Modern medicine arose when scientists learned to fight some of the worst infectious diseases with vaccines and drugs. This strategy has not worked with AIDS, malaria, and a range of other diseases because of their complexity and the way they infiltrate processes in cells. Curing such infectious diseases, cancer, and the health problems that arise from defective genes will require a new type of medicine based on a thorough understanding of how cells work and the development of new methods to manipulate what happens inside them.

Today's research is painting a picture of life that is much richer and more complex than anyone imagined just a few decades ago. Modern science has given us new insights into human nature that bring along a great many questions and many new responsibilities. Discoveries are being made at an amazing pace, but they usually concern tiny details of biochemistry or the functions of networks of molecules within cells that are hard to explain in headlines or short newspaper articles. So the communication gap between the worlds of research, schools, and the public is widening at the worst possible time. In the near future young people will be called on to make decisions—large political ones and very personal ones—about how science is practiced and how its findings are applied. Should there be limits on research into stem cells or other types of human cells? What kinds of diagnostic tests should be performed on embryos or children? How should information about a person's genes be used? How can privacy be protected in an age when everyone carries a readout of his or her personal genome on a memory card? These questions will be difficult to answer, and decisions should not be made without a good understanding of the issues.

I was largely unaware of this amazing scientific revolution until 12 years ago, when I was hired to create a public information office at one of the world's most renowned research laboratories. Since that time I have had the great privilege of working alongside some of today's greatest researchers, talking to them daily, writing about their work, and picking their brains about the world that today's science is creating. These books aim to share those experiences with the young people who will shape tomorrow's science and live in the world that it makes possible.

Acknowledgments

This book would not have been possible without the help of many people. First, I want to thank the dozens of scientists with whom I have worked over the past 12 years, who have spent a great amount of time introducing me to the world of molecular biology. In particular, I thank Volker Wiersdorff, Patricia Kahn, Eric Karsenti, Thomas Graf, Nadia Rosenthal, and Walter Birchmeier. My agent, Jodie Rhodes, was instrumental in planning and launching the project. Frank Darmstadt, executive editor, kept things on track and made great contributions to the quality of the text. Sincere thanks go as well to the Production and Art Departments for their invaluable contributions. I am very grateful to Beth Oakes for locating the photographs for the entire set. Finally, I thank my family for all their support. That begins with my parents, Ed and Jo Hodge, who somehow figured out how to raise a young writer, and extends to my wife and children, who are still learning how to live with one.

Introduction

As a white blood cell roams through the body on the trail of bacteria and other invaders, it resembles an amoeba searching for food in a mossy pond. But, the amoeba leads an independent life, whereas the blood cell is a citizen of a vast apparatus—a human being. Working with more than 50 trillion *clones* of itself, it builds and operates the body, sometimes for a century or more, before the whole structure falls apart in death. The theme of this book is how that enormous living machine arises seemingly by itself, but actually as a dialogue between each cell's *genes* and the environment.

From one point of view this process begins anew with every child, but most of the biology of building a human body can be traced back to events that happened early in the evolution of animals. Key developmental genes were already in place in a small, wormlike animal that lived more than half a billion years ago and became the direct ancestor of all animal life on Earth today. The blueprint of its body was so successful that all of its descendants—including human beings—have inherited the basic plan virtually intact.

Developmental biology is the branch of science devoted to the great scientific mystery of how plants and animals arise from single cells. Until modern times these processes could only be investigated by watching visible changes in embryos and seedlings. The invention of high-resolution microscopes in the 19th century brought a dramatic leap forward with the discovery that animals were made of trillions of cells, each of which was born through the division of another. From that moment on, answering developmental questions would require understanding processes within cells.

In the 1950s James Watson (1928–) and Francis Crick (1916–2004) discovered *DNA*'s role in heredity, evolution, and the life of the cell, opening a new way forward. The blueprint of development had now been located; it was encoded in the genes in each cell's nucleus. Cells read and interpreted it to form the piston-like building blocks of muscle, the sprawling, treelike shapes of neurons, and all of the body's other types of cells and tissues.

One aim of developmental biology is to learn how a healthy body arises from an organism's *genome*. Another goal is to understand why development sometimes goes wrong. *Mutations* may produce organisms with harmless (or potentially beneficial) characteristics such as extra fingers. At other times they lead to fatal developmental defects or problems that make it impossible for an organism to lead a healthy life. They are also the reason that new species can arise. Over the past few years research has exposed connections between these questions that have drawn together development, medicine, and several other branches of biology into a new way of thinking about organisms. Many people are still not familiar with these latest insights. This book aims to explain what is happening at a level that high school students and other interested nonspecialists can understand, without any special prior knowledge.

Today thousands of researchers in laboratories across the world are trying to figure out how genes guide specific developmental processes in humans and other animals. The field is so large that a book of this size can only capture a small part. The basic principles of development remain nearly the same across animal species and throughout a lifetime, but following the later stages requires some detailed knowledge of anatomy. This volume will therefore focus mainly on the early stages in humans, showing how they have been studied in animals that are closely related to our species through evolution. It will introduce the major molecules and genetic processes that occur as an embryo grows and will show how they shape tissues and organs.

As an example, one of the most interesting parts of developmental biology concerns the origins of organs in the very early embryo. Long before the body forms recognizable parts, cells receive instructions that determine their fates. Sometimes an entire organ can be traced back to a key event involving a single molecule. For example, if a gene called cerberus is not activated in a certain set of cells at just the right time, an embryo will not develop a head or brain. What tells these cells to switch on the cerberus gene, and what prevents other cells from switching it on? This is the kind of question that interests molecular biologists, and today they have developed fascinating new methods to answer it.

Chapter 1 introduces the unicellular stage of a human life from its origins as separate sperm and egg cells. Those cells arise from much older cells, created shortly after our own parents' birth, and they have a long history leading up to the moment of fertilization. Chapter 2 takes up the story just after fertilization, following the growth of a few identical cells to a stage at which there are many types, organized in a very primitive body structure. Chapter 3 follows the next steps in which cells form all of the embryo's major organs and a body that is recognizably human.

Chapter 4 is devoted to the relationship between diseases and cell differentiation—one of the most important topics in current biomedical research. Every day researchers learn more about the intimate connection between problems such as cancer and the normal genetic programs that transform *stem cells* into specialized daughters such as white blood cells or the glial cells of the brain. A thorough understanding of healthy and unhealthy cell differentiation may soon allow researchers to transform specialized cells back into potent stem cells that can be used in therapies.

Finally, chapter 5 shows how development and evolution are coming together to show how changes in just a few genes can create new species that are remarkably different from one another. Comparisons between living species are allowing scientists to solve some evolutionary puzzles such as the origins of complex organs like the eye. There are also some practical applications of the intersection between these fields: Researchers can develop much better animal models of human diseases.

Developmental biology has had an important social impact by explaining the origin of many developmental defects that inspired fear and superstition in the past. Today we know that even people with severe developmental defects may be almost completely normal from the genetic point of view. Their problems may arise from the change of a single letter in the billions that make up the genetic code. The completion of the human genome sequencing project has made it much easier to pinpoint the source of developmental defects and genetic diseases.

Knowledge helps us empathize with those who are not so lucky and accept their differences. Chapter 3 describes a problem called *microcephaly* in which a mutation leads to children who are born with a tiny brain. The birth of such a child often poses huge problems for families, but sometimes it can be uplifting as well. While doing the research for this book, my family and I ate at a restaurant in central Missouri whose owners had such a child. The baby lived only a few months, but the story of her short life was captured in photos and letters on a bulletin board for all the customers to see. It was obvious that even in that short time the family had showered the infant with love. This family surely would have reacted the same way whether they knew the reasons for the unique development of their child or not. For the rest of us, however, learning a bit about development may help us react with compassion and understanding rather than fear.

1

Life as a Single Cell: Creating a New Animal

Everyday soil is home to one of the strangest organisms on Earth, a species of slime mold called *Dictyostelium* whose lifestyle is more bizarre than most creatures from science fiction. In times of plenty it lives as a single-celled amoeba, but when the food supply grows short, it has a unique strategy for survival. Tens of thousands of independent cells come together and build themselves into a slug-like creature that can sense changes in light and temperature. It crawls along the ground to a new location where it stops and changes shape, becoming a plantlike stalk topped by a bulbous head. Eventually the head bursts, scattering spores that create new organisms. *Dictyostelium* lives in a strange niche between single-celled and multicellular life. Scientists are studying the organism to learn more about how animals evolved and how cells manage the job of building bodies.

Before animals could evolve, cells had to develop modes of communication that would help them act in a coordinated way. Such systems still underlie the contraction of muscles and cross talk between brain cells. Cells also had to cling to each other, but not permanently; they had to be able to let go to migrate, form layers, and fold. Even though *Dictyostelium* is only distantly related to humans and mammals, it has made a good model organism to study some of these basic processes in animal cells. In 1967 Theo

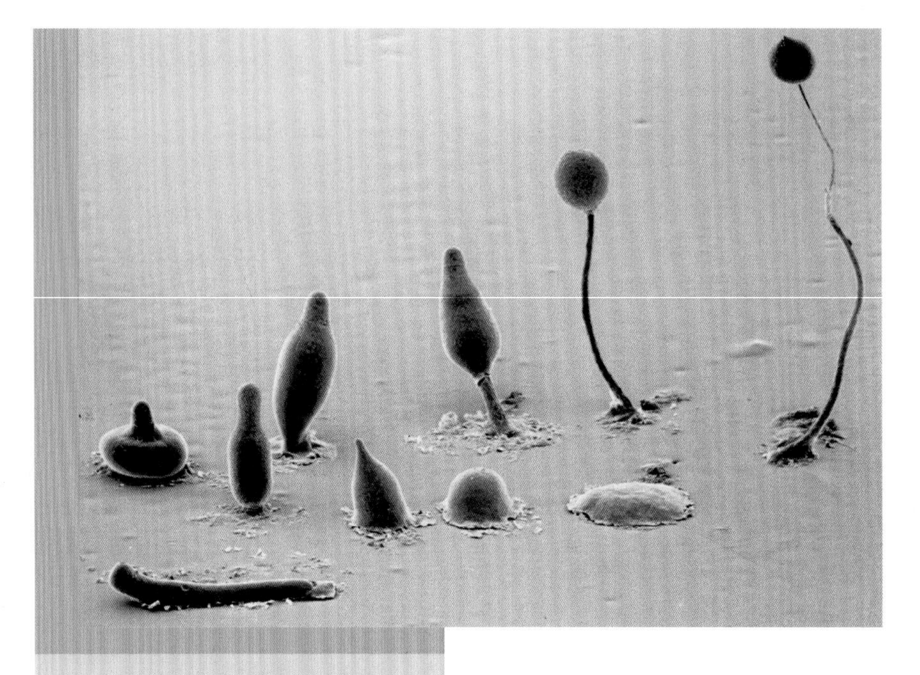

Counterclockwise from lower right: When the single-celled *Dictyostelium* lacks food, it clusters with others to form a small, worm-shaped animal that crawls to a new environment. There it forms a stalk-like structure with a bulbous head that eventually bursts, spreading cells that once again live an independent life. *(M. J. Grimson and R. L. Blanton, Biological Sciences Electron Microscopy Laboratory, Texas Tech University)*

Kunijn at Princeton University and David Barkley, working with colleagues in Utrecht in the Netherlands, discovered the signaling molecules that guide individual *Dictyostelium* cells as they form clusters. This substance, called AMP, is also used by human white blood cells as a navigation signal. Ten years later Günther Gerisch and Kurt Müller, working at the University of Basel, discovered the first *protein* responsible for helping the amoeba stick to its cousins. This molecule also has relatives in human cells that help hold tissues together.

Life on Earth began as a single cell, and each human alive today began as one fertilized egg. In both cases, becoming a true animal involves a number of complex and dramatic steps. After a brief look at the oldest forms of multicellular life, this chapter describes the first phases of a human life—the events that occur just before and just after fertilization—which are crucial to what happens later.

MAKING BODIES FROM CELLS

Unicellular species may have existed for more than 2 billion years before the evolution of fungi, plants, and animals. *Dictyostelium* has already provided some clues as to how this great evolutionary step happened. Insights have also been gained through the study of an ocean organism called the *choanoflagellate.* This single-celled organism is shaped like a basket. At one end is a hairlike flagellum that beats like a propeller, drawing water and food particles into the basket, where they can be absorbed. Sometimes hundreds or thousands of choanoflagellates band together in colonies; for example, the ball-shaped organism called volvox is made up of hundreds of them. Their flagella face outward on the surface of the ball and can spin the colony or move it through the water. Volvox meets one of the characteristics considered essential for a true multicellular organism: It has more than one cell type. (In fact, it only has two. A few of its cells take on the task of reproduction, and the rest lose that ability.) Volvox lacks, however, a second defining feature: It does not really undergo a process of development.

Scientists such as Roel Nusse of Stanford University have been focusing on a gene called Wnt as a key influence on the development of multicellular life. Wnt is found in all complex organisms but not in single-celled creatures. Nusse's lab has conducted experiments to watch the effects of switching Wnt on and off in animal cells. Recent work shows that the molecule plays a key role in whether cells stick to one another or are released to migrate. In 2006 Nusse and his colleagues showed that when the Wnt gene is activated in fly embryos, it temporarily loosens the connections between cells and allows them to migrate. It is equally crucial in determining whether stem cells continue to divide or specialize. One molecule therefore combines several functions that are important for development.

Defects in Wnt or cellular molecules that help interpret the signal have also been linked to cancer, in which the same processes are often disrupted.

Choanoflagellates caught the attention of early evolutionists who hoped to understand the origins of plant and animal life. Evolution worked in small steps, so there ought to have been intermediate forms between single cells and animal life. Choanoflagellates seemed to fit the bill because of their relationship to the sponge. These ancient animals come in many shapes and forms, but they share some common features: an outer surface of pores through which water enters, an inner compartment that absorbs food, and a large opening through which water is pushed out again. The interior is lined by cells called *choanocytes* that look and behave almost exactly like choanoflagellates. By beating their flagella, they draw water in through the pores and provide food for the sponge.

Choanocytes help the sponge solve the engineering problem of delivering food. Each cell of a simple creature such as volvox comes into direct contact with the environment and can extract what it needs. This also works in organisms that are just a few layers thick. Enough food and water can seep through or be passed along from one cell to the next. And, the same is true of human embryos at a very early stage, but a growing organism quickly becomes too thick to supply oxygen and food to cells on the inside. The solution for the sponge is its network of pores and its inner bag; for humans it is a stomach and system of blood vessels. The circulation of very small organisms does not have to be very complex, but greater size requires a pump—the heart—that can push nutrients to the far regions of the body.

Size and shape pose other structural problems that have to be solved with each phase of an organism's growth. Sometimes cells have to be woven tightly together into tissues; at other times they are released to migrate, fold, and form new organs. Most sensory organs have to be on or just below the surface, but they also have to remain connected to an inner brain as the skin or shell expands and carries them away. Just as cities expand and engineers have to build roads and lay electrical and water lines, organisms constantly remodel their nervous and circulatory systems to keep up with the amazing pace of growth. All of these biological construction projects are self-organizing and are carried out by single cells—but without a centralized "mission control" to oversee how things are going and make course corrections. Success ultimately depends on cells receiving information about their locations and differentiating in the right way.

This process is fairly simple in volvox, which has only two cell types, but creating a human requires the development of hundreds of specialized cell types out of more generic stem cells. The details will be discussed in chapter 4, but many insights into the process have come from simple organisms. When they live on their own, the cells of *Dictyostelium* are identical, but as they come together, they take on many distinct roles. Some, for example, become light sensors, some are responsible for helping the sluglike animal move around, and others control its growth into a plantlike stalk.

Experiments that interfere with single molecules have revealed genes that are required for each of these transformations, yet it was difficult to get an overall picture until the completion of *Dictyostelium*'s genome in 2005, which was carried out by an international consortium of 22 laboratories. They discovered that the organism has approximately 12,500 genes—nearly half of the number that have been found in the human genome and only about 2,000 fewer than the number scientists have found in the fruit fly. That gives it about twice the number found in yeast cells, which also lie on the evolutionary branch leading to animals but which spend their lives as unicellular organisms. Somewhere in this extra code are the keys to multicellularity. Many of the genes necessary for *Dictyostelium*'s animal-like behavior have been passed down throughout the animal kingdom, where they guide a wide range of developmental processes.

THE RISE OF SEXUAL REPRODUCTION

As multicellular organisms evolved, so did a new way of creating them: sexual reproduction. Previously, new organisms had been created through cell division, and interestingly, aside from the moment of fertilization, a human being is still the product of this process that spawned its ancient ancestors. Once egg and sperm have fused, the new cell divides over and over again to produce the body.

Most life on Earth still reproduces asexually; in other words, an organism does not have to find a partner in order to reproduce, and the process usually creates clones—copies—of the parents. But, once sexual reproduction evolved, it offered advantages to some species' survival and evolution. Charles Darwin (1809–82) pointed out that the amount of diversity within a species often had an important influence on whether it survived or became extinct. If all the individuals in a population were nearly identical, they would be affected the same ways when the climate grew colder, or a new predator arrived, or something else about the environment changed.

Each organism produced by sexual reproduction, however, has an original recipe—a unique mixture of genes from its parents. Any differences, even small ones, may give a particular individual an advantage at surviving and passing along its genes. If the advantages are also passed along to the offspring, the latter also go on to have more offspring, and this process goes on long enough, the overall makeup of a species changes. With sufficient time the result may be a new species. This bias in reproduction is called *natural selection,* and it is a driving force in evolution. Species with large populations and a lot of variation usually survive much longer than those with little variety because there is more likely to be an individual somewhere that can live and reproduce if conditions change.

Most new plants and animals are conceived when a sperm provided by the father (pollen in plants) fuses with an egg (also called an *ovum,* or an *oocyte*) from the mother, but this is not always the case. Many species of fish, reptiles, amphibians, and ticks reproduce through *parthenogenesis.* Here an egg develops without being fertilized by sperm; one organism plays the role of both father and mother. Some of these species go through the motions of mating anyway. This type of "play" is even necessary sometimes to start the development of the egg, although scientists do not yet know why. Whiptail lizards are an interesting example. They are all females, but when it comes time to reproduce, one animal has to play the role of a male. David Crews, a zoologist at the University of Texas in Austin, has been studying the lizards since he discovered their unusual mating behavior in the 1970s. His recent work on the genetics of whiptails reveals that their sexual roles are assigned by two hormones. If a lizard produces a large amount of progesterone, it will play the male role, while estrogen-producing animals act as females.

Lizards and other animals conceived this way may not be absolutely identical, because their DNA also undergoes mutations, but such species show less variety than do organisms that create the next generation through sex. Diversity comes not only from the fact that receiving genes from two parents mixes them up in new ways but also from the unique way that egg and sperm cells are created.

THE BEGINNINGS OF A LIFE: EGG AND SPERM

A look at an organism's germ cells shows how crucial they are to the survival of a species, and as such they are protected carefully within the body. Germ cells are familiar to most people because of their role in reproduction. But, getting them ready to create a new organism is a lifelong process that begins when the parents themselves are embryos, shortly after their own conception.

Germ cells are among the first types to arise in a developing embryo, and they are carefully protected from specializing, which is the fate of nearly every other cell. They usually have to migrate great distances to get to sex organs, which form much later as the body grows. They find their way by reading cues on other cells that attract or repulse them, much like the way single *Dictyostelium* cells find each other by tasting chemical signals in the environment.

Germ cells keep dividing to produce young, vital cells and remain immature until a person reaches puberty. One step in preparing them for reproduction is to split up their DNA into two half-sets. That is necessary because the fertilized egg (and the future cells of the new person) is a *diploid*: There are two copies of each *chromosome,* one inherited from each parent. That is also true of germ cells for most of their lifetimes. They divide over and over again to create a stock of *precursor cells* that are put on hold, like pressing a developmental "pause" button.

When a woman is born, her ovary holds about a million primary oocytes that are kept for long periods of time—up to 50 years. Not all the cells survive; by the time the woman reaches puberty, about 200,000 will be left. Since the 1950s scientists believed that all of the oocytes a woman could ever produce were made before birth, but a study in 2004 by Jonathan Tilly, director of the Vincent Center for Reproductive Biology of Massachusetts General Hospital, showed that women have stem cells that can make new ones. Tilly and his colleagues discovered that the oocytes of newborn mice die so quickly that if they were not replaced, the mice would have none left when it came time to reproduce. This meant that there had to be another source for oocytes, and Tilly's genetic studies proved that new ones could arise from stem cells lying on the surface of the ovaries. The same thing may happen in humans, and studying the process may reveal why the production of eggs stops when a woman reaches menopause, the end of the fertile phase of her life.

At sexual maturity females produce hormones that prompt the development of eggs. Each cycle, which comes about every 29.5 days, causes half a dozen or more oocytes to begin to mature. A compartment called the *follicle* forms around each egg. Its job is to nourish the egg, which grows until it undergoes *meiosis,* a special type of cell division that removes the second chromosome of each pair.

After another week the woman's body releases other hormones that prompt *ovulation.* During this phase the oocyte travels into a passageway called the oviduct and on through the *fallopian tubes.* On its way the follicle breaks apart, leaving the egg and some of its accompanying cells. The egg dies within about a day unless it is fertilized, which happens in the upper region of the oviduct. The fertilized egg travels to the uterus, becomes

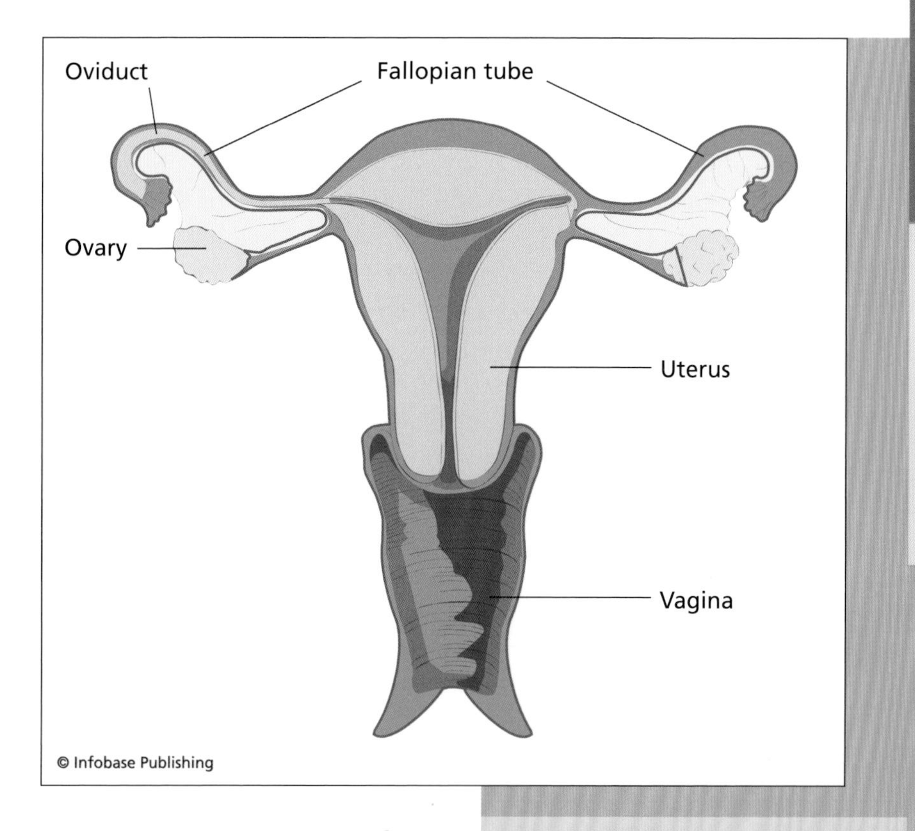

Each egg matures inside a cluster of cells called a follicle, in the ovary. During ovulation a follicle moves into the oviduct, where the egg is released. If it is fertilized it moves on to the fallopian tubes and the uterus.

implanted there, and if everything goes well it forms an embryo.

In men, a single germ cell produces four sperm in a process that lasts about 65 days. Sperm are produced constantly, unlike in females where eggs start to mature in a monthly rhythm. The process takes place in a male's testicles, which produce about 200 million sperm each day. First the germ cell's DNA is copied, leading to a large structure called a spermatid that contains four copies of each chromosome. Soon afterward this is broken down into individual cells that undergo dramatic changes of shape and form. The nucleus shrinks to become a small "head," and most of the other regions of the cell are lost. Proteins snap together into long tubelike structures that form a tail—a flagellum—which allows the sperm to swim toward the egg. Cellular "power plants" called *mitochondria,* located between the head and the tail, provide the flagellum with energy.

Walter Sutton (1877–1916) of the University of Kansas and Nettie Stevens (1861–1912) of Bryn Mawr College, one of the few American women to receive a Ph.D. in science at the turn of the 20th century, figured out the role that chromosomes played in determining an embryo's sex. In humans and most animals the sperm determines whether an embryo will become male or female; eggs can become either sex. The reason lies in the 23 pairs of chromosomes in human cells. In both men and women 22 of these are matched pairs: The chromosomes are identical in size and shape, and they contain the same genes. In females this is also true of the 23rd pair, which contains two matched *X chromosomes.* But, in males the 23rd pair consists of one X chromosome and a much smaller *Y chromosome.* As the female's 23rd pair consists of two Xs, the splitting of the pair gives each of her eggs an X. Because males have an XY pair, splitting it up leaves half the sperm with an X and the other half with a Y. Depending on which of these fuses with the egg, the result is either an XX or XY pair, which determines the embryo's sex. Thus, from the moment of creation each sperm is destined to become a particular sex, providing it can fuse with an egg.

This is the normal case, but studies of humans have revealed exceptions: A small proportion of women have only one X chromosome, and a few males have two Xs plus a Y. This means that something about the Y makes an embryo into a male (rather than the fact that one X is missing), but what? At the beginning of the 1980s scientists still had no idea. Then a series of studies by Robin Lovell-Badge, a geneticist at the National Institute for Medical Research, with Anne McLaren and Paul Burgoyne of the Medical Research Council (both in London) discovered that a gene called SRY, normally found on the Y chromosome, had to be present for mice to become male. Soon they showed that SRY plays a similar role in humans. In very rare cases SRY "jumps" onto an X chromosome and is found in embryos with two Xs, turning what would normally be a female into a male. (The reasons for this odd behavior are explained in the sidebar "The Odd Case of Jumping Genes.") At other times the egg inherits the XY pair, but the SRY gene is missing, which leads to a female embryo.

The discoveries had a completely unexpected consequence. For several years the International Olympic Committee had carried out medical inspections to ensure that the participants in women's events were really females. Now it seemed the SRY gene could be turned into a genetic test for gender, and it was put to use for the 1996 Summer Olympics in Atlanta, Georgia. Several females tested positive for the SRY gene, but the test was flawed and no one was barred from competing. After a protest by several U.S. medical associations, genetic testing for gender was dropped again in 2000. In the meantime, other genes on the Y chromosome have been found to contribute to the development of male sex characteristics.

The earliest sexually reproducing animals were probably *hermaphrodites,* animals that have the reproductive organs of both sexes; most plants are like this. There are many theories about how two distinct sexes and the Y chromosome evolved. A popular theory holds that the Y began as a second X that shrank and lost genes. Over the past few hundred million years, genes on the Y chromosome have evolved specialized functions having to do with reproduction and the development of male sex characteristics. In some animals the loss of Y genes has become extreme. The kangaroo Y chromosome contains only one gene. Male grasshoppers have no Y at all; they have a single X, where females have two.

Nature has produced many other ways by which an embryo's sex is determined. The situation is particularly bizarre in the platypus, one of very few mammals that lays eggs. It has 26 pairs of chromosomes, five of which are XY in the male. As with the male grasshopper, some other insects have a single X chromosome, with no Y at all. In birds the male has identical chromosomes, while females have one unmatched pair. In other species one sex receives a set with three chromosomes, and the other, only two. Some creatures, like the whiptail lizards

The Odd Case of Jumping Genes

One reason that the SRY gene sometimes "goes missing" is that a child does not simply inherit whole chromosomes intact from his or her parents. After a careful study of the chromosomes of corn under the microscope, geneticist Barbara McClintock (1902–92) made the radical claim that genes could change positions, get turned around backward, or even hop to another chromosome. Few people believed her at the time; it took decades for the mainstream to accept the concept of a "jumping gene." The long wait paid off: In 1983, at the age of 81, McClintock was finally recognized with a Nobel Prize for this and a lifetime of other discoveries in genetics. Normally the prizes in each category are shared by two or three scientists; McClintock was the first woman ever to receive the Nobel Prize in physiology or medicine alone.

Her work followed some other important studies of chromosomes. While watching strandlike chromosomes line up side by side before their separation, a Belgian researcher named Frans Alfons Janssens (1863–1924) saw that they were twisted in sharp bends. Thomas Hunt Morgan (1866–1945), another Nobel Prize–winning geneticist of the early 20th century, believed that the twists could create so much pressure they might break the two

discussed in the previous section, switch between being male or female at different phases of their lives.

Then there are cases like alligators. In 1982 Mark Ferguson of the Queen's University of Belfast, Ireland, teamed up with Ted Joanen of the Louisiana Wildlife and Fisheries Commission to solve a puzzle. Male and female alligators have identical chromosomes; so why are there two sexes? The two researchers chromosomes at parallel sites. Depending on where the breaks occurred, this would leave fragments of various sizes, from partial genes to segments containing several genes. (This is what McClintock observed in her microscope studies of corn.) Cells could repair the damage, but they might not be able to tell exactly where a fragment came from. The process of repair might reinsert it in the wrong place—even in the wrong chromosome—in a process called *recombination*. Sometimes pieces were put in backward (*transposition*). This would not necessarily cause serious problems for an organism, but it influences how that bit of DNA is inherited and could change the way genes work. Recombination is regarded as a main cause of changes in genomes, creating variety that sometimes leads to new species.

Barbara McClintock on November 7, 1983 *(Barbara McClintock Papers, American Philosophical Society)*

proved that the answer lies in the temperature at which the egg is incubated. Between the first and third weeks after conception, temperatures below 86 degrees Fahrenheit (30°C) cause all eggs to develop into females; higher than 93°F (34°C) produces males. Sex in many species of lizards and most turtles is also determined in this way. This means that a female can influence the sex of her future offspring by where she lays her eggs.

In species such as humans, germ cells are programmed for a particular sex by the chromosomes and their genes. It is possible that other aspects of the body are preprogrammed through an egg's proteins. The oocyte of the fruit fly, *Drosophila melanogaster,* develops a "rough draft" of the fly body even before fertilization. This unfertilized egg is vast, like an unsettled continent that has not been divided into states. It is a rich storehouse full of proteins and other molecules that nourish the embryo through its first stages of development. After fertilization the nucleus divides over and over again, but the egg does not get any larger and the new nuclei are not packed into individual cells until about a thousand copies have been made. (The situation is similar to a married couple who moves into one bedroom of a huge mansion; as their children, grandchildren, and future generations are born, they move into the other bedrooms without having to build on to the house.)

Before fertilization molecules enter the fly oocyte from the mother's body. They are placed in precise locations that act as a coordinate system that tells the major parts of the body where to grow. Just as regions of the United States have different resources—mines and minerals, fertile soil for growing crops, lakes and coastlines rich in fish—molecules in the egg create regions with specialties. So each of the new cells finds itself in a region with its own chemistry, and this stimulates it to develop in a particular ways. For example, only one edge holds the "oskar" protein. Cells that arise in this region respond to the protein by building the posterior (hind) end of the fly. If oskar is produced elsewhere, by mistake, cells become confused. The fly will never develop an abdomen, which is obviously fatal. The position of the bicoid protein defines where the head and thorax will develop. Other proteins have similar effects on other regions of the body. These molecules do not act alone; they depend on helpers to put them in the right places, and many questions remain about why they have their effects on development.

A pioneer in this field is the German researcher Christiane Nüsslein-Volhard (1942–   ), who shared the Nobel Prize in 1995 with Eric Wieschaus (1947–   ) and Edward Lewis (1918–2004). The three geneticists' work on the fruit fly in the late 1970s and

The body plan of a fly is influenced by events that take place even before fertilization of the oocyte (the large oval compartment on the right). A protein called oskar (red) can only be produced at the posterior end. If it is produced elsewhere, the fly will not develop an abdomen.

1980s created a new kind of developmental biology that was strongly tied to molecular biology. When Nüsslein-Volhard and Wieschaus arrived at the European Molecular Biology Laboratory in Germany in 1979, they decided to try to identify genes from the mother fly that influenced the development of embryos. To create mutations they fed male flies sugar water containing substances that damaged DNA. Then they allowed them to mate with females. This often produced malformed embryos, and the scientists could work backward to discover which genes had which effects. The work required Nüsslein-Volhard and Wieschaus to spend several months peering at embryos under a microscope with two sets of eyepieces, looking for developmental defects.

Initially the plan was greeted skeptically, even by their bosses. At the beginning of the 20th century the fly had been the star of genetics research—Morgan's lab had used the insect to identify hundreds of new genes—but most of that work had involved adult flies, and it was now regarded as old-fashioned biology. Yet the project paid off quickly. The fly turned out to be ideal for the new project. Its embryo passes through an early stage in which its body is divided into obvious segments. In some of the mutants these stripelike regions were obviously disturbed: They were in the wrong place or completely missing. Since each segment gives rise to major body structures, the mutations revealed genes that were crucial to the development of flies and most other animals. Those genes appear over and over throughout the rest of this book.

In mammals, genes in the fertilized egg become active quickly to take control of development. Many of these are related to molecules that guide early developmental processes in insects and even more remote evolutionary relatives. Yet, there is a difference: Genes do not become active in most insects until the fertilized egg has divided for many generations; molecules from the mother are crucial in getting things started in the right direction. There is some evidence that the placement of maternal molecules also influences human development, but the importance of this process is not yet clear.

FERTILIZATION

During the act of sex (sexual intercourse), a man releases about 200 million sperm. If the conditions are right, one—and only one—may fertilize an egg. During intercourse the sperm enters the female reproductive tract, then has to make a long swim to the upper part of the oviduct, near the ovaries, where the egg is located. The fusion of a sperm and the egg is necessary to begin a new life.

Locating the egg is a challenge very much like the one *Dictyostelium* cells face when they cluster to begin building an animal, discussed at the beginning of the chapter. In some species it is solved in a similar way: The egg releases molecules that sperm sense and follow. For animals that mate in water the attraction has to be especially powerful to keep sperm from drifting away. Their eggs release powerful chemical signals that sperm follow in a process called *chemotaxis.* The easiest way to think of this is to imagine trying to find the source of a smell by sniffing the air. The nose is sensitive to both the type and the concentration of certain molecules, and the surface of the sperm bears proteins known as receptors that behave in a similar way. The small directional molecules dock onto the receptors, triggering changes in the behavior of the tail and the direction of the swimmer.

Signaling molecules and receptors fit each other much in the way of a very sophisticated key and lock, which solves another problem in water. There may be many species of sperm in the water, and they have to be attracted to eggs of the right species. (Here, too, there is a similarity to smell; cells in the nose have many types of receptors, each of which is unlocked by specific chemical "keys.") Only molecules released by the right species of egg fit the lock of a matching sperm. Different molecules trigger different types of swimming behavior. In sea urchins and many other species, molecules close to the egg have a dramatic effect on a sperm: They switch motors in the tail into high gear, giving it a final burst of speed as it approaches the egg.

The guidance system works a bit differently in humans. Muscles in the female uterus create waves that propel sperm most of the way to their goal in the upper oviduct. Signals released by the egg probably play a role at close range, but sperm also rely on cues from the cells that line the inner passageways, like constantly stopping to ask for directions along the way. Within about 30 minutes, an amazingly short time given the distance to be covered, some of the sperm reach the oviduct, but these cells are not likely to fertilize the egg. Studies in 1995 by Allen Wilcox and Donna Baird, of the National Institute of Environmental Health Sciences in North Carolina, studied the timing of conception and found that sperm can survive six days in order to fertilize an egg. The earliest to arrive might not be the most successful. Studies by other groups suggest that sperm have to undergo a round of preparation, called "capacitation," when they enter the female reproductive tract. Scientists do not yet know just how long this takes in humans or exactly what it involves. There may only be a small window of opportunity around the time of ovulation in which it has to happen.

When a sperm has finally arrived, it still has to penetrate a protective layer around the egg. In insects and other invertebrate animals this is the *vitelline envelope,* a mixture of cells and proteins that helps ensure the egg and sperm are of the same species. The envelope is wrapped in a second, jelly-like layer of protection. Human and mammalian eggs are surrounded by the *zona pellucida,* a thick region of proteins and some of the nurse cells that grew alongside the egg inside the follicles. Those cells accompany it during its release and entry into the oviduct.

A protein called ZP3 helps a sperm bind to the zona pellucida. Next the sperm has to cut its way through the zone using *enzymes,* proteins that digest other molecules. These are packed into the sperm's head and become activated when it butts its way into the zone. The enzymes permit it to penetrate the layer and finally come into contact with the outer membrane of the egg.

Once in contact the two cells must fuse their membranes, like two soap bubbles that meet and meld into one. The nucleus, mitochondria, and tail of the sperm enter the main compartment of the egg, the *cytoplasm,* in a process that is also guided by proteins on the cells' surfaces. In humans a specific area on the side of the head of the sperm has to come into contact with the egg membrane, probably because of specific proteins located there.

Entry into the egg is followed by an immediate reaction that prevents other sperm from following; otherwise, the egg might collect three or more copies of each chromosome. The egg protects itself in several ways. First, the docking proteins used to tie onto the first sperm are thrown away. Then the egg releases substances that make it much harder for other sperm to penetrate the zona pellucida. Finally, there are major changes to the membrane enveloping the cell. One of its functions is to keep the cell's electrical charge in balance with that of the environment by taking in and releasing charged particles. An unfertilized egg has a strong negative charge because it exports a lot of positively charged particles; for a reason still unknown, this helps sperm fuse and enter. The entry of a sperm triggers a reaction that flashes across the membrane, causing the cell to release calcium. This dramatically changes the cell's charge and repulses any more sperm that are trying to enter.

The successful sperm is destroyed except for its nucleus. In particular, the mitochondria must be destroyed because the egg has its own. Mitochondria have their own DNA; they lie outside the nucleus and copy themselves independently, almost like parasites that have infected the cell. In fact, according to Lynn Margulis, a biologist at Boston University, this is how they originally evolved. In the 1970s she proposed that they began as independent organisms, probably small bacteria, that took up residence in an ancient cell and established a mutually beneficial relationship. Recent comparisons of the DNA of mitochondria with bacteria seem to confirm this hypothesis. In any case only the mother's mitochondria remain in the fertilized egg. Because this DNA is passed from mother to daughter, it has provided a way to build family trees for the female side of the human race far back into evolutionary history.

The DNA of the sperm and egg have to come together to merge. This is managed by protein towing lines called *microtubules,* a network normally used to deliver molecules through the cell. Microtubules grow or shrink by adding or removing small protein subunits, much like making a tower of Legos. In the egg, multiple "towers" are built outward from a small structure, a centriole, to form a star-shaped pattern called an "aster." Some of the microtubules encounter the nuclei of the sperm and the egg, attach themselves, and then reel them in close enough to fuse. Gerald Schatten's group at the Oregon Regional Primate Research Center has helped show that infertility in some men arises because centrioles cannot make the aster pattern.

In frogs and many other species the point of entry of the sperm is important to what happens next. A sperm can enter from any direction, but once it has, the microtubule network has to be rebuilt to bring its nucleus to that of the egg. This creates a sort of highway through the cell, an axis that will determine which parts of the egg become which parts of the future body.

Fusion of the cells' DNA creates a cell (at this point called a *zygote*) that once again possesses a full set of chromosomes. It now has everything it needs to create a complete animal. The new set of genes becomes active, turning out *RNA*s that are translated into proteins, and soon the cell will begin to make copies of itself.

In rare cases a mother releases two eggs and both become fertilized. If the zygotes survive and become implanted in her uterus (at this point becoming embryos), they will develop into *fraternal twins.* Each of the embryos has a unique set of DNA, so they are as different from each other as any brother or sister. *Identical twins,* discussed in the next section, originate from a single fertilized egg.

CLEAVAGE AND THE CELL CYCLE

The "molecular age" began in 1953 when the American researcher James Watson and his British colleague Francis Crick discovered that DNA molecules in the cell nucleus contain genes, the units of heredity and the instructions for building and operating the body. For the discovery they were awarded the 1962 Nobel Prize in physiology or medicine, along with the physicist Maurice Wilkins (1916–2004). Since then, explaining the development of organisms has meant understanding why there are changes in the activity of genes and how they affect the behavior of cells and the formation of tissues. These changes are usually triggered by proteins, so the challenge is to discover which protein switches on a particular gene, and what tells it to do so.

Until fertilization, most business within the egg is handled by molecules that are produced by the mother's cells and then brought inside. In many species these maternal molecules continue to direct the most important events after fertilization, but in mammals the new embryo almost immediately takes an important step toward arranging its own affairs. The DNA in its cells is used to produce new RNAs and proteins, the worker molecules responsible for most of the cell's big projects. Proteins manage cell division, transmit information from the environment to the genes, change the cell's chemistry, give the cell its shape, and carry out thousands of other tasks. Activating new genes alters the structure and behavior of the cell. Differences in the molecules produced by different types of cells result in the creation of specialized cells, tissues, and organs.

Two main things happen between fertilization and the formation of the first recognizable organs. First, the zygote divides over and over again in a process called *cleavage*. This happens very rapidly and produces a large number of cells called *blastomeres*. They wrap themselves into the shape of a hollow ball, a *blastula,* creating inside and outside compartments. The second major development, *gastrulation,* is covered in chapter 2.

Cleavage is the stage at which identical twins can form. If some blastomeres become separated from each other at an early stage, they may develop into separate embryos. Because they started as the same egg, they have identical DNA. Their bodies will develop in nearly identical ways, although not perfectly identical because development is constantly influenced by subtle events in the environment, and many of an organism's features are not determined by genes. For example, every human embryo is programmed to develop fingerprints, but the precise patterns they form are determined by tiny, random events that vary from individual to individual.

Sometimes at a very early stage the twins fuse again. If they are not identical, this creates a human *chimera,* a single individual with two complete sets of genes. A chimera's body develops as a mixture between the sets. The first case of a human chimera was discovered during a blood test taken in 1953. A subject known as "Mrs. McK" gave blood at a clinic in northern England, and doctors were astounded to discover that she seemed to contain two blood types. Robert Race, an investigator at the Medical Research Council in London, had heard of such a case occurring in twin cows and wondered if Mrs. McK had a twin. In fact, she did have a twin brother who died at three years of age, initially suggesting that their blood might have somehow been "mixed" during development. Actually, however, the blood probably came from a triplet, whose cells had fused with Mrs. McK's very early in development.

In the meantime, more than 30 other human chimeras have been found. In most cases evidence of a second set of genes has been found only in the person's blood. (When a blood test seemed to show that Tyler Hamilton, an American cyclist who competed in the 2004 Tour de France, had undergone blood

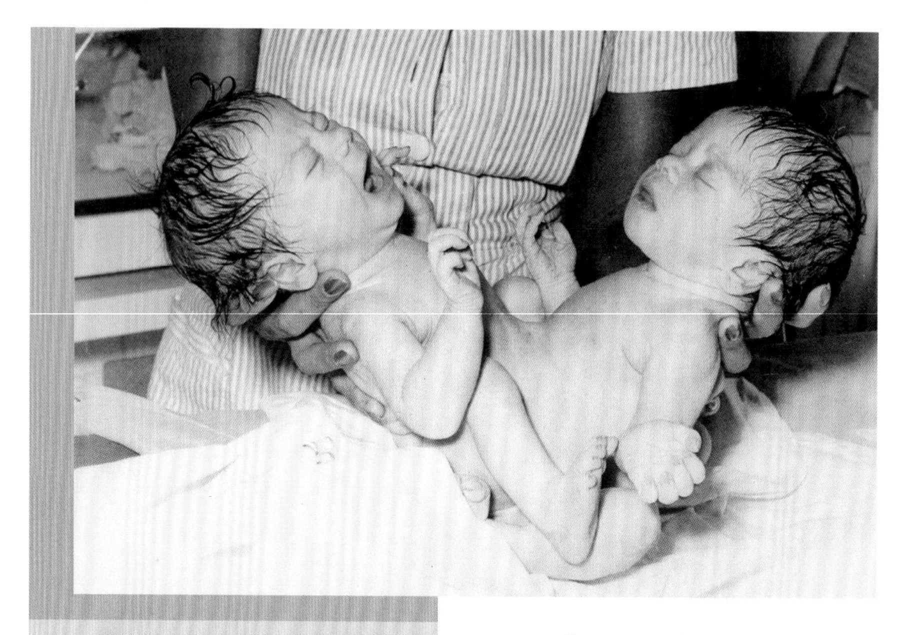

The conjoined twins Clara and Alta Rodriguez were separated successfully in 1974 by Dr. C. Everett Koop. *(National Library of Medicine, National Institutes of Health)*

transfusions during the race, he countered by claiming that he was a chimera. After numerous additional tests the claim was not accepted by the Court of Arbitration for Sport.) In some cases more than blood is affected; the entire body is a patchwork of two genomes, combined in unusual ways.

From a blood test to see if "Jane" could donate a kidney, this woman learned of her chimera condition. The results showed that only one of her three sons could be her genetic child—until further tests revealed that her body contained traces of a vanished twin whose cells had produced the eggs of the other two boys. Margot Kruskall, who worked on the case in 2003 at the Beth Israel Deaconess Medical Center in Boston, concluded that Jane was the product of twin girls who fused as zygotes. Cases have also been discovered in which the twins were of different sexes, sometimes leading to individuals who developed characteristics of both males and females.

Other irregularities during cleavage can have a huge impact on development. One result is conjoined twins (also known as

"Siamese twins," after the brothers Chang and Eng Bunker, who lived in the 19th century and became famous as part of P. T. Barnum's circus), whose bodies are joined at a certain part. Scientists believe that such cases happen when the egg loses its orientation during early cleavage. Instead of dividing along one axis—the way that a book falls open on a table—cells form two axes. This creates two "centers of development" that lead to duplicate body parts in some regions. Since the discovery of chimeras, some scientists have proposed that conjoined twins may also happen when twin blastulas fuse at a late stage.

Cleavage sets up patterns of growth that will determine the entire future of the embryo. Cells in some parts of the egg reproduce faster than others, following a sort of internal clock whose pace is set by the production of certain molecules. In the very early embryo the *cell cycle* is fairly simple: Cells alternate between phases of rest and reproduction. Later they need to increase their size before reproducing, and two stages of expansion are added to the cycle. The phases in this process are shown in the figure on p. 24. Particular proteins push cells from one phase of the cycle to the next.

While all species begin to grow through cell divisions, the "growth maps" of species can be quite different. The first cells of mammalian zygotes divide right down the middle, but the zygotes of amphibians, birds, reptiles, and many other organisms make asymmetrical cells. This has to do with the amount and placement of *yolk* within the egg. Most people know yolk as the mass of yellow material in chicken eggs, which is made mostly of proteins that are produced in the mother's liver and brought into the egg before fertilization. In birds the yolk is obvious because it takes up most of the egg. But the cells of other animals also contain yolk, and the amount and its position in the egg have an important influence on cleavage.

Charting what happens during cleavage is like following population maps of the United States over decades of census taking. Growth in some areas is quicker than in others. Cells divide slowly in regions with a lot of yolk; where there is little, the process happens much faster. In fish and birds only one small area to the side of the egg is free of yolk, and cleavage

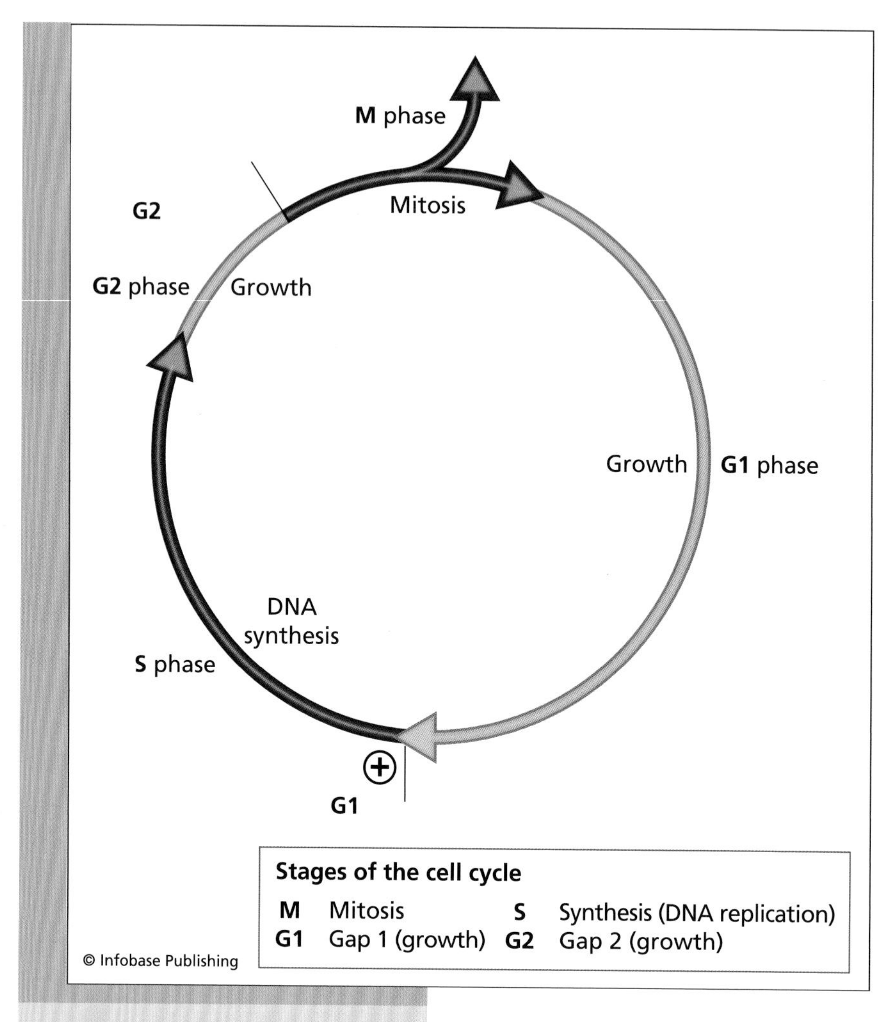

The stages of a cell cycle

happens in this small area. In insects the yolk occupies most of the center of the egg, so cells divide around the rim. Because yolk is spread thinly and evenly through the eggs of mammals, the differences are not as extreme. Later, other factors influence the rate of division, and imbalances arise between the regions.

(opposite) The amount and location of yolk in the egg and the orientation of the mitotic spindle define where and how quickly cells divide in the early embryo. This is important in establishing an animal's basic body axes.

THE MACHINES OF CELL DIVISION

One of the architectural wonders of nature is a structure that builds itself inside cells to carry out cell division, the *mitotic spindle,* which is made of microtubules. Microtubules act as struts and scaffolds in the *cytoskeleton,* a system of protein fibers that give cells their shape. They also serve as a sort of subway,

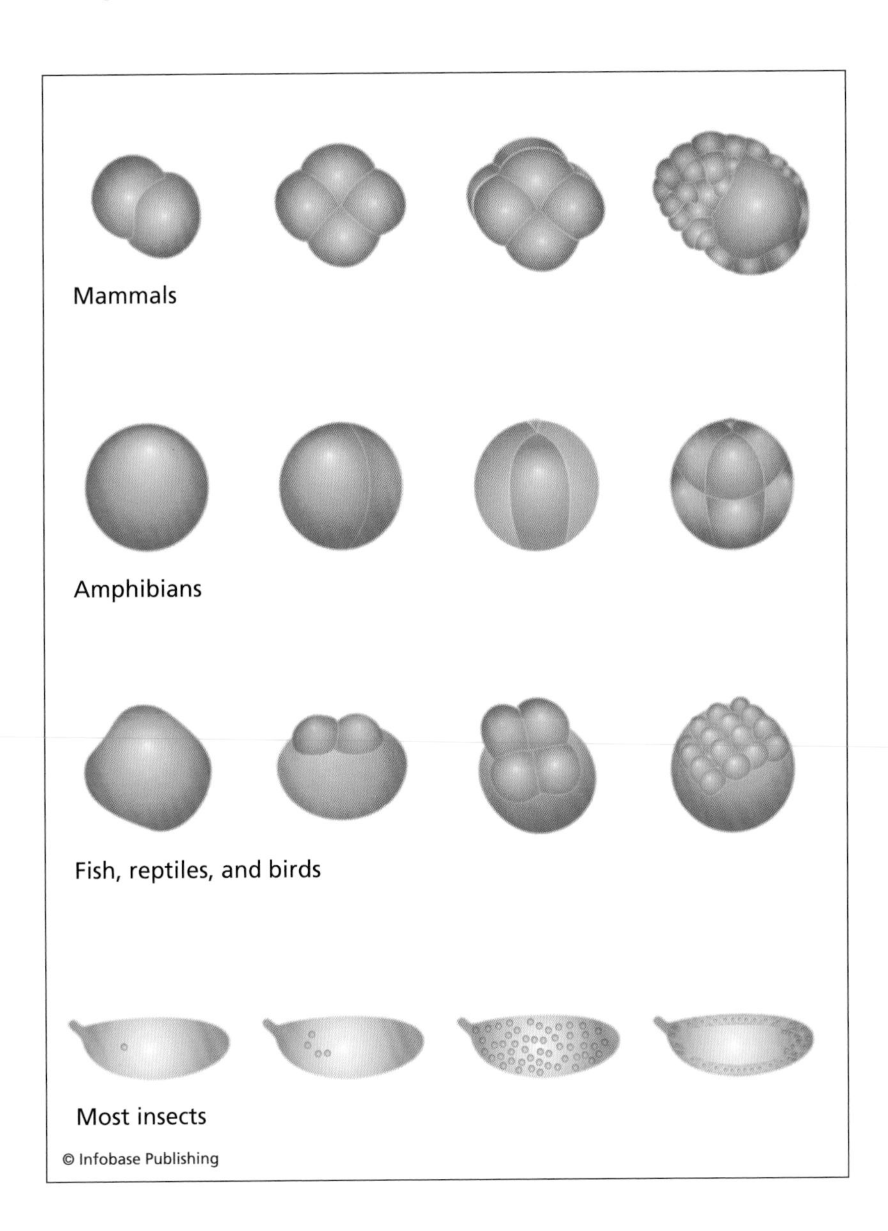

shuttling material through the cell (although here the passengers—proteins and other molecules—run along the outside rather than inside the tube). They are built of protein subunits that stack onto one another, like using Legos to build a hollow tube. Usually construction starts at a small structure called a *centrosome,* located near the nucleus. From this point microtubules spread throughout the cell in a tangled network, a bit like holding the strings of hundreds of kites in one hand and reeling them out into the sky in all directions.

This whole system is disassembled and rebuilt during cell division—but first a word about what happens to the nucleus. The membrane around the old nucleus is dismantled. Some of the smaller pieces are left connected to each other and shuttled out into other cell membranes, like leaving blocks of puzzle pieces together when it is returned to its box. Later, the pieces will be recalled and new ones added, but they will be split up to contribute to the assembly of two fresh nuclei.

Now the centrosome copies itself, and one copy wanders to the opposite side of the nucleus. This creates two kite flyers, standing on opposite sides of the cell. Microtubule "strings" begin stretching toward the center, where chromosomes have been lined up in a flat disk. At the same time microtubules have begun building themselves outward from the chromosomes. They meet up with the fibers from the centrosomes and tie themselves together. When the process is complete, the centrosomes begin reeling in chromosomes, drawing them to opposite sides of the cell.

Eric Karsenti, a cell biologist at the European Molecular Biology Laboratory in Heidelberg, Germany, has been investigating the mitotic spindle for more than 20 years, trying to understand how microtubules self-organize to create this beautiful structure. Many researchers have focused their attention on the centrosomes, but Karsenti knew that this was not the complete story. Sometimes spindles formed without any centrosomes at all, because microtubules could also be built outward from DNA. He developed a method to re-create this effect in the test tube using extracts from dividing frog cells. By adding and removing molecules from this mixture and modeling their physical properties in the computer, he could test how various proteins affected the spindle.

In 2005 Karsenti teamed up with microscopist Philippe Bastiaens to try to solve a riddle: Why did centrosomes produce short microtubules that shrank and grew at a fast rate, yet those built outward from DNA were much longer and more stable? The scientists presented their answer in an article published in the journal *Science.* A combination of two proteins creates a special chemistry around the nucleus, which changes farther away from DNA. Where there is a high concentration of the two molecules, microtubules become long and stable. Farther out, where the centrosomes are located, one of the proteins is bound to a third molecule. Here the microtubules are shorter and more dynamic. Many other proteins are needed to govern other aspects of the spindle's shape and behavior, such as whether microtubules successfully capture chromosomes and manage to pull them apart.

Mitotic spindle. The mitotic spindle is built of fibers called microtubules, which reorganize into a spindle shape to separate chromosomes during cell division. *(MIC Bergen)*

Now the cell has to be pinched off to form two new complete cells. This process, called *cytokinesis,* is a bit like pinching a balloon to divide it into two separate bulbs of air. Usually this involves the creation of new membrane (because two new cells have a greater surface area than the single parent), built from internal cell membranes, newly synthesized fats, and other raw materials. The microtubule network made to create the spindle has to be reconfigured as a delivery route for some of these components. Bits of membrane are delivered along with proteins needed to direct the pinching behavior. Otherwise, the two projects would get out of sync with each other.

Cell division can produce either two identical cells—such as the first cells that are created in most zygotes—or asymmetric cells. Which one happens depends on the behavior of the spindle and cytokinesis. If centrosomes take mirror-image positions opposite each other, the result is usually symmetrical division. But one centrosome may stay in the middle of the cell while the other migrates to a border, setting up the geography to make one large and one small cell.

Most specialized cells arise when a generic stem cell divides asymmetrically, leaving one cell that will differentiate into a specific type and one that will remain a stem cell like the parent. The specialized cell goes on to take a job within a tissue, and it usually stops reproducing, whereas the stem cell is kept in reserve so that more "specialists" can be made when the time comes.

Asymmetries and the direction of early cell divisions determine the future body orientation of an animal. The beautiful spiral shape of snails originates from the angles of the mitotic spindle (relative to the bulk of the yolk), the fact that each cleavage produces one large and one small cell, and the positions that these cells take in the body. When four small cells have been made, they are moved to the left or right of their larger partners. This happens over and over again, creating a turning pattern. Whether the pattern is clockwise or counter-clockwise depends on a single event: the angle of the mitotic spindle at the four-cell stage.

If something goes wrong in the cell division machinery or in the signals that control the cell cycle, the results are almost always disastrous. In the first stages of development this will likely disrupt the entire body; later problems may affect single organs or cause cancer. Cells in the early embryo reproduce at an amazing pace that slows down considerably by birth and drops for the rest of a person's lifetime. Most cells become specialized and lose their ability to reproduce. In some types of cancer, however, specialized cells reactivate genetic programs that make them behave like an embryonic cell. The links between stem cells and cancer are discussed in detail in chapter 4.

SUMMARY

With some interesting exceptions animal lives begin as separate germ cells in the bodies of male and female parents. These cells are set aside very early in the parent's life; they have to be protected from developing into specialized cells so that they retain the capacity to produce an entire organism. Egg and sperm undergo a special type of development of their own which prepares them for fertilization when the parents mate. In some species this preparation helps set up coordinates for the future body plan of the embryo.

Comparing these processes in different species sheds light on how they evolved and how multicellular life arose in the first place. For ethical and practical reasons much of what is known about human reproduction started off in studies of fish and other organisms. The evolutionary relationships of organisms mean that related genes carry out many of the same functions, in spite of "superficial" differences. These themes and their implications for medicine and other fields are explored in depth in chapter 5.

2

The Rise of Structure: Early Growth and Differentiation

What makes a human being different from a colony of cells such as a volvox? The first stages of life, in which cells divide and form a ball-like shape, look superficially similar, but volvox never develops further, whereas a human embryo continues to grow and form increasingly specialized tissues. Fertilization and cleavage produce a few cells that are very similar, but they soon begin to take on individual characteristics. Chapter 1 introduced the main reason this happens: Each type of cell activates a unique subset of its entire collection of genes. This changes the cell's collection of molecules, its structure, and the way it behaves. Most of the changes are originally triggered by stimulation from other cells or the environment. They lead to gastrulation, the next step in building a body, which involves mass migrations of cells. This process sets the stage for the formation of organs and is the main topic of this chapter.

CLEAVAGE AND IMPLANTATION OF THE EMBRYO

At the end of cleavage most animals consist of a ball-shaped *blastocyst,* made up of blastomeres. The precise shape of the embryo varies from species to species because of the position of yolk in the original egg. Cells divide more slowly in the region containing the most yolk, called the *vegetal hemisphere.* In the rest, the *animal hemisphere,* things proceed at an incredible pace. In less than 48 hours a frog egg divides to create nearly 40,000 cells. In flies the process is even faster; 50,000 cells are produced in the first 12 hours of life. (Later, things slow down.) In mammals cleavage happens much slower. It takes 1.5 days for a fertilized mouse egg to undergo its first division, and in 3.5 days there are only 32 cells. Later, cell division will speed up, although it will never match the early pace of a fly.

By the 32-cell stage, mammalian cells have shaped themselves into a blastocyst; in the center lies a liquid-filled cavity called the *blastocoel,* which will play an important role as cells begin to migrate. In frogs the blastocoel is a large, inland sea; in humans it is an elongated strip, like the slit on the head of a screw.

The first chapter described how it takes several days for a human zygote to travel from the site of fertilization to its home for the next nine months, the uterus, where it becomes an embryo. The first rounds of cleavage happen during the voyage. The zygote is moved along by hair-like cilia that line the oviduct. It is still enclosed in the zona pellucida, the protective envelope that helps attract sperm to the egg and then repel them (described in chapter 1). Now it has another function; it keeps the zygote from attaching itself to the oviduct on its way to the uterus.

Despite the protection, sometimes the zygote begins to grow in the oviduct as an ectopic (out-of-place) pregnancy. While the new embryo often simply stops developing and its cells disband, in other cases the situation poses a serious threat to the mother's health. The oviduct can become infected and rupture, causing pain and dangerous bleeding. Ectopic pregnancies can usually be treated by medications or surgery, but they still may be life threatening, especially if the mother does not have access to good medical facilities. Ectopic pregnancies are rare, but the rate has been rising. Until about the 1970s they accounted for fewer than 1 percent of all pregnancies in the United States, but

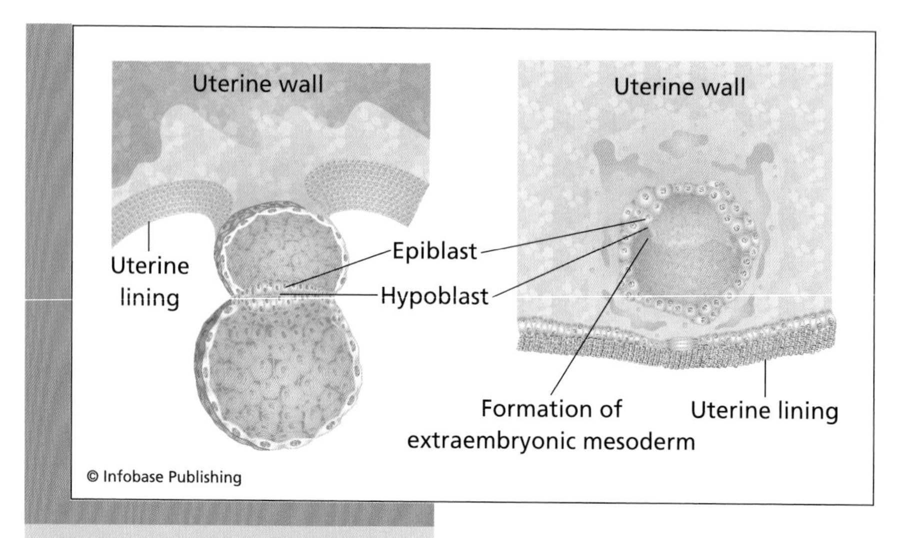

When the blastocyst arrives in the uterus, its enzymes eat through the uterine lining, and the blastocyst is embedded in the uterine wall, where it will take its next developmental steps.

by 1980 the rate had doubled. This number also varies in different regions of the world; in Jamaica, for example, about 3.6 percent of pregnancies are ectopic. The reasons for the differences and changes are unknown.

If everything goes normally, however, the zygote enters the uterus about six days after fertilization. First, it "hatches" by boring a hole in the zona pellucida. The zygote emerges and bumps against the surface of the uterus, where proteins tie it up tightly. At this stage it consists of 32 cells. These are still nearly identical, but they have formed inner and outer layers that have a crucial role to play in the next steps of development. Ultimately, the outer layer, called the "trophoblast," will no longer belong to the embryo at all; it will be used to build a bridge to the mother's body.

As the layers switch on different sets of genes, the trophoblast goes to work on the wall of the uterus by secreting enzymes that dissolve its lining. The new embryo squeezes through the gap and embeds itself in the underlying tissue, called the uterine wall. As the trophoblast carves out space for the embryo, it triggers the mother's body to send new blood vessels into the area and sends out fingerlike extensions to create more contacts with the mother's tissue and thus to receive more blood. These connections remain and will eventually develop into the umbilical cord, the main connection to the mother's body, a pipeline that feeds the growing child. Another job of the trophoblast is to "negotiate" with the mother's immune system. The embryo is a new organism with new genes; without this help the body might mount an immune reaction and reject it, the way transplanted organs are sometimes rejected.

Other trophoblast cells develop into the placenta, a sac that protects the embryo, nourishes it, and disposes of waste. The placenta is filled with a liquid called the amniotic fluid. As the embryo grows, it sheds some of its cells into the placenta. By withdrawing some of this fluid in a procedure called "amniocentesis," a doctor can obtain a sample of the cells, grow them in laboratory cultures, and then examine their genes and chromosomes for signs of severe deformities or genetic diseases. (A look at the 23rd chromosome pair also reveals whether the embryo is male or female.)

Amniocentesis is increasingly being replaced by a new technique that is safer for both the mother and embryo because it can be performed earlier in the pregnancy. Amniocentesis can only be performed after the 14th week of life, and results are only obtained after the cells have grown in the laboratory for about 14 days. This may not leave much time to act if the embryo is developing in an abnormal way that might threaten the life of the mother. The newer method involves sampling cells that will form the placenta. They also arise from the egg, so they are equally useful in learning about the embryo's genes.

GASTRULATION: LAYERS AND MIGRATIONS

After cleavage comes the process of gastrulation, which British biologist Lewis Wolpert has called "the most important time in your life." Cells continue to divide at a rapid pace around the blastocoel. This quickly produces three layers that take on unique characteristics. They slide along each other and fold into each other. Each migration brings cells into contact with new neighbors that speak a new chemical language. This process of *induction* gives them instructions that activate genes and tell them what to become.

The three layers are the *ectoderm* (outer layer), *mesoderm* (middle layer), and *endoderm* (inner layer). They will produce hundreds of specific cell types and the body's organs. (The exception is germ cells, which have already been set aside.) Some organs arise from a single layer; others are built from the cells of two or three layers. Even when one layer is the source of an entire organ—the brain develops from the ectoderm, for example—it cannot happen without input from the other layers.

Gastrulation was discovered in the early 19th century. While studying the embryos of chicks, the Russian scientist Christian Pander (1794–1865) discovered the three layers and understood that they had to interact to produce organs. He could not dig more deeply into the means by which this happened and soon gave up biology to become a paleontologist. The work was taken up by his countryman Karl Ernst von Baer (1792–1876), the greatest embryologist of the early 19th century, who noticed that the embryos of different species went through remarkably similar stages during early development. This discovery later helped bring the fields of evolution and development together and is discussed in chapter 5.

The major branches of life undergo gastrulation in different ways. In invertebrate animals such as sea urchins, the ectoderm, mesoderm, and endoderm form as a result of sheets that grow and fold. After they have formed a ring of cells, a few cells migrate into the interior (the blastocoel) from the slow-growing side of the embryo—the yolky, vegetal side. Then the blastula dimples inward. This is quite different than the process in flies, whose eggs are full of yolk. There, the first cells form a ring around its huge blastocoel. Gastrulation finally begins when about 1,000 cells fold inward to create a furrow that stretches partway around the embryo's midline, like a belt only large enough to pass halfway around. As they push

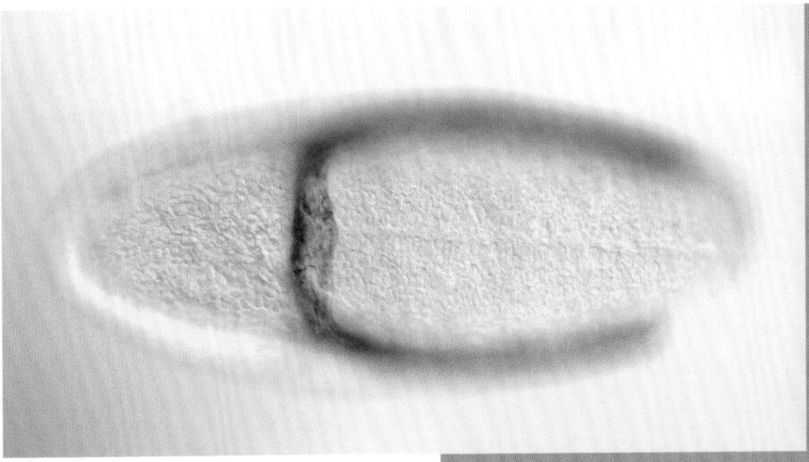

Gastrulation. During gastrulation, cells form three layers that go on to shape themselves into the body's tissues and organs. *(Lawrence Berkeley National Laboratory)*

inward, they form an internal loop, the way a paper folds when its edges are pushed toward each other. With further growth the fold becomes a tube. The creases where the tube is pinched off detach and form a new layer that will become the mesoderm and endoderm. There is still an indentation on the surface where the fold began; it is now covered by cells that grow and migrate from the sides.

Human and mammal eggs have little yolk, so gastrulation happens in yet another way. The outer layer becomes structures to feed and hold the embryo in the mother. Inside the blastula dividing cells first create two layers: The *epiblast* covers a lower, thinner layer of cells, the *hypoblast.* Most of the future body comes from the epiblast. Its cells divide and thicken into a central region called the *primitive streak.* This grows and stretches into a long band that will become the head-to-tail axis as the animal grows.

At this point the process becomes hard to understand without thinking in three dimensions, and it will be helpful to introduce the "navigation system" scientists have created to describe the embryo. The axis from left to right is easy; it is just called left to right (from the perspective of the organism itself).

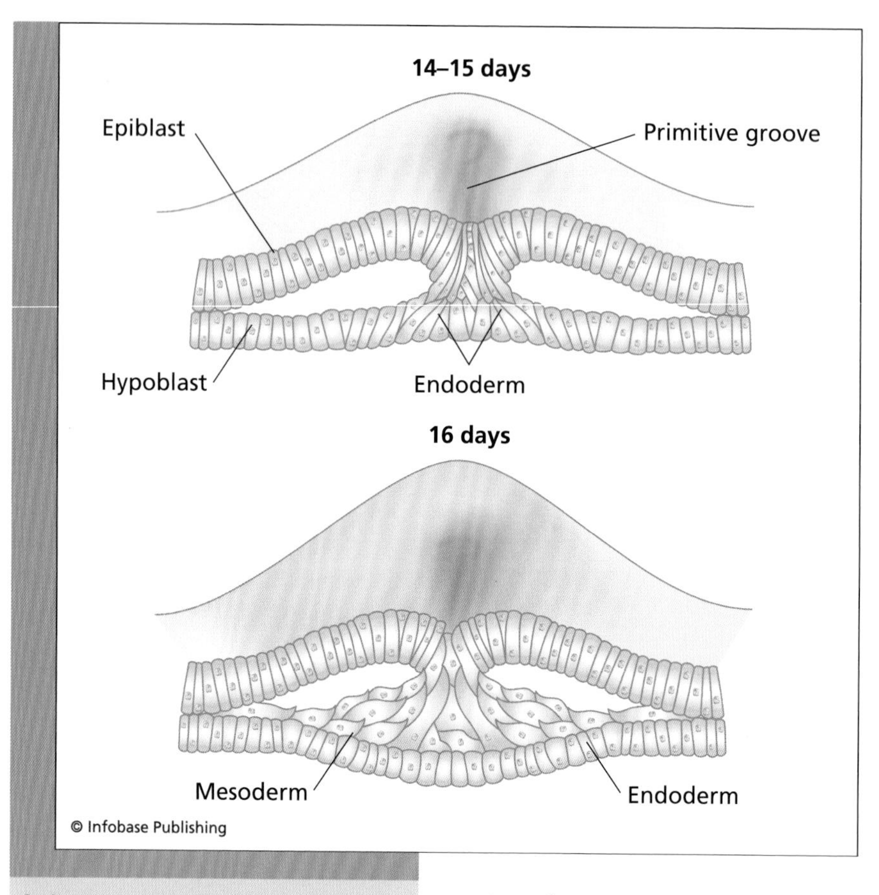

In humans gastrulation begins when cells from the epiblast migrate downward until they reach the hypoblast. This transforms two layers into three: The epiblast becomes ectoderm, the migrating cells become mesoderm, and the hypoblast becomes endoderm.

Another axis runs from the head (anterior) to the abdomen (posterior); the legs are not yet important. Front is called "ventral," and the back is "dorsal." These terms will be used throughout the rest of the book.

With these coordinates in mind, what does the embryo look like after the formation of the primitive streak? It is oval shaped, slightly longer in the anterior-posterior (up-to-down) direction. The primitive streak runs up and down, just under the dorsal surface (the back), like a sort of spine.

At a key location called the node (the site of the future head), cells cast off their ties from their neighbors and migrate downward through the primitive streak. Once they arrive at the hypoblast, they move to the sides, pushing aside the cells that are already there. The newcomers will develop into red blood cells and the mesodermal organs, such as the heart and muscles. The hypoblast will transform itself into the endoderm and will develop into the stomach, intestine, lungs, and other endodormal organs. The last cells that come down from the node form into a tubelike shape called the *notochord,* which will develop into the nervous system and brain. The epiblast cells left behind will become the ectoderm, which later becomes organs such as the skin, glands, and parts of the eye.

ORGANIZERS: THE EXPERIMENTS OF SPEMANN AND MANGOLD

Many of the basic principles of gastrulation and their importance in development became clear through the experiments of the German scientist Hans Spemann (1869–1941) and his graduate student Hilde Mangold (1898–1924) in the early 20th century. Spemann received the 1935 Nobel Prize in medicine and physiology for their work. (Mangold died in a tragic explosion at her home in 1924.) Their main method was to perform "microsurgery" on the embryos of frogs and other animals, removing bits of tissue, reimplanting them in new places, or transplanting them between embryos to watch their effects on development.

Spemann's early work followed up on experiments by Wilhelm Roux (1850–1924) and Hans Driesch (1867–1941), who split early animal embryos into single cells to try to discover for which parts of the body these cells were responsible. Up to a certain point in development each blastomere could develop into an entire organism. This was especially puzzling to scientists because they did not yet know where genes were located or what they were made of or that each cell of the body carried a complete set of hereditary information. Studying the fruit fly, Morgan's lab at Columbia University was making great progress in discovering new genes, but this work was almost entirely focused on traits that appeared in adults. It would be many decades before genetics and developmental biology were brought together through the work of Nüsslein-Volhard and Wieschaus (described in chapter 1).

Not all ways of dividing up embryonic cells led to complete animal bodies. Sometimes removing one of the first two cells in a frog led to an animal with only half a body. This led to an important question: How did cells know what to become? Did each cell contain specific instructions from the beginning, a sort of preprogramming, or was it flexible and able to learn through communication with its neighbors? Transplanting cells ought to give an answer. If a blastomere's fate was predetermined, then it would develop in a certain way even when moved to the wrong place. If it obtained instructions from neighbors, then the new neighborhood would override the programming of a cell when it was moved.

Spemann's first experiments showed that splitting an early newt embryo in different ways had different developmental outcomes. Critical was a small zone called the "gray crescent," which formed on the surface of the zygote after fertilization. The simplest way to think of this might be to imagine an orange with a gray spot running horizontally along its peel. (The slices of the orange divide it up the way cells form within the space of the egg.) Just as there are many ways to cut an orange in half, there are several ways to divide the cells of the embryo. If Spemann split them so that each half contained part of the gray crescent, the result was two halves that grew into complete, healthy newts. That was not the case after a cut that left the gray crescent entirely on one side; then only that half would develop into a newt. The other would grow into a mass of undifferentiated tissue.

Scientists had already learned to track the fates of early cells as they grew, differentiated, and moved by marking single blastomeres with dyes. Spemann and Mangold took advantage of this to understand what was happening in the embryos. They started by taking cells that would become epidermis and then moving them to the neural plate, and vice versa. These embryos developed normally, as if nothing had happened at all. Clearly the cells they had transplanted were still flexible and depended on instructions from their neighbors to develop.

The researchers now began systematic studies to find out if the entire embryo followed the same pattern. The gray crescent did not. This area forms a small fold, called the blastopore lip, and is the site at which cells begin their migration from the outer layer inward. When Mangold and Spemann removed the lip and transplanted it to a new embryo, it still followed its old program. The recipient embryo developed two complete sets of nervous systems—two spines and two heads. Because the transplants caused a reorganization of their neighbors, Spemann called this region an "organizer." It was so potent that a small number of cells taken from it could redirect the development of a much larger structure.

Further work showed that organisms contained several more organizers. If a portion of the upper part of the blastopore lip was transplanted, it created a second brain in the trunk region of another embryo. This led to a series of famous images of two-headed embryos that can now be seen in nearly every biology textbook.

Spemann expanded these studies to address another question: What made similar species different? He transplanted tissues between the embryos of closely related animals. Cells from one type of newt could form gills in another, but very interestingly, the gills resembled those of the donor species rather than the recipient. An even more dramatic experiment produced tadpole mouth structures within newts and vice-versa.

Spemann concluded that a cell's surroundings do not completely guide development; instead, their job is to activate a cell's own developmental programming through chemical signals. He began trying to find those substances, without success;

How Signals Work

A key part of developmental biology is devoted to understanding how cells respond to the environment and one another, and the basis of those responses is chemistry. The first task is to identify the molecules that are responsible for a job, then to find out which other molecules they work with and how they affect genes.

Signaling usually begins with a *ligand,* a small molecule that is either released by a cell or sits on its surface. The ligand docks onto a receptor protein on a second cell. There are many types of receptors; in a typical case the molecule floats in the cell membrane, with a head region extending outward, a part that passes through the membrane, and a tail inside the cell cytoplasm. Ligands often dock onto more than one receptor and draw them together. This brings two or more tails together, which wrap around each other and cause changes in shape and chemistry. The result is a small "chain reaction" in which new proteins come along, dock onto the tails, and are themselves changed. They usually go on to influence many more molecules. At some point the signal reaches a *transcription factor* that can enter the nucleus. It docks onto DNA, usually in pairs or in combination with other molecules, and changes the status of nearby genes. Often it does this by providing a docking place for a large molecular machine that transcribes the gene into RNA, or it may remove molecules that have been blocking transcription.

A very common chemical "language" of signaling is based on transferring clusters of atoms called "phosphate

at the time too little was understood about the chemistry of the cell. Finding answers would take several more decades, but Spemann's work had revealed a number of fundamental principles groups" from one protein to another. This process, *phosphorylation,* uses one of the cellular "energy currencies," such as *adenosine triphosphate,* or ATP. This very simple molecule is built of one of the bases found in DNA (adenosine), attached to a sugar and three phosphate groups. It switches between a high-energy state in which it is loaded with the three groups and a lower-energy state in which one of them has been stripped away. This back-and-forth switching is managed by other molecules, which borrow the energy to drive the cell's chemical processes.

Some of the body's most important signals work in a much more direct way. Hormones, for example, are so small that they may slip through the cell membrane without a receptor. Some of them are whisked through the cell directly to the nucleus, where they act as transcription factors or have other effects on genes.

Signals are crucial to everything that happens in the cell and the organism as a whole. One job is to tell cells when to reproduce (numerous cancers have been linked to defects in signaling molecules; they sometimes get stuck in the "on" position and tell the cell to divide all the time). Another is to respond to changes in the environment, for example, to activate new sets of genes when the temperature rises or when food gets scarce. Signals also inform cells of their positions in the body, telling them how to shape themselves into tissues and organs and guiding the process of development. Biochemical signals are the closest that development comes to having an architect, yet there are many local bosses, rather than a single main planner at the top.

of animal development. A reenactment of the experiments can be seen on the Web site of the journal *Nature* (see Further Reading at the end of this book).

MOLECULAR SIGNALS IN GASTRULATION

Organizers are clearly special, but what makes them that way? Researchers realized that location was important. Usually the organizer developed along the boundary between slow-dividing cells in the vegetal area and fast dividers in the animal side. The borderline created an axis; one side became the belly and front of the animal (ventral), the other the dorsal side. Even in mammalian embryos, which had no yolk to speak of, animal and vegetal zones developed and created this main body axis. Would the organizer function properly from another position in the egg?

In the 1980s biologist John Gerhart and his laboratory at the University of California at Berkeley carried out an interesting set of experiments to find answers to some of these questions. In frogs the initial entry point of the sperm into the egg determines the location of the organizer. After the sperm has entered, the cell's membrane makes a small turn toward this point, like turning a globe so that the east coast rather than the west coast of the United States faces the front. Somehow this triggers the beginning of gastrulation on the opposite side of the globe—about the position of East Africa.

First, Gerhart held the eggs in place so that they could not turn. This caused severe problems: The dorsal side of the frog, along with structures such as the vertebrae and the brain, did not develop. The embryo contained endoderm, ectoderm, and mesoderm, but no major organs developed, leaving a sort of belly without the rest of the body. Somehow the cell's rotation was important in activating the dorsal organizer.

In another experiment Gerhart and his team allowed the membrane to shift but then made an extra turn of the fertilized egg. If the sperm entry point was turned to the top, the cytoplasm adjusted so that gastrulation began at the top. The most interesting findings came, however, when they allowed the normal rotation to happen and then gave the cell an extra spin, away from the sperm entry point. This led to the development of two organizers (one at the normal position and another at the entry point of the sperm) and tadpoles with two heads, two brains, two backs that joined at the trunk, and a single tail. The scientists could get the same results by taking just three cells from the organizer region and transplanting them into another embryo. From their new location they could instruct neighboring cells to start building a back. This was clear evidence that the organizer worked by sending out developmental signals to other cells. What molecules were being used, and what information did they transmit?

A Dutch embryologist, Pieter Nieuwkoop (1917–96), and his lab in the Netherlands made a major breakthrough in answering these questions, first by removing and transplanting tissues in much the way Spemann had. But more than 50 years had passed since Spemann's work, the genetic code had been cracked, and now there was a bigger goal. Scientists hoped not only to say where signals came from but to identify the signaling proteins and their effects on genes in other cells.

Nieuwkoop removed cells from the mid-regions of embryos—those that normally developed into mesoderm. He discovered that acting on its own, neither the endoderm nor the ectoderm could prompt other cells to become a mesoderm. However, if the outer and inner layers came into contact, cells from the fast-growing ectoderm were able to create tissues such as muscles, blood, and nervous system structures that normally are built from the mesoderm. Further work pinned down this power to the activity of a few cells on the dorsal side, now known as the Nieuwkoop center. Thus the development of the mesoderm is stimulated by signals between early endoderm and ectoderm layers.

The next step was to find out which molecules were involved. Studies by Janet Heasman at the University of Minnesota and Kathleen Guger of Memorial Sloan-Kettering Cancer Center in New York revealed that a protein called beta-catenin might be the signal or at least played a role in passing it along. After fertilization this molecule accumulated on the dorsal side of the embryo, and in later stages it appeared in the cells of the Nieuwkoop center. Beta-catenin is a transcription factor: It combines with other proteins to dock onto genes. In doing so, it changes the chemistry and behavior of cells.

Since this discovery, researchers have found that beta-catenin is involved in a wide range of developmental processes and is also frequently involved in cancer. It is so important that a number of other molecules are involved in controlling when and where it can become active. Some of these trigger it to become active; others keep it on hold. Understanding how beta-catenin functions meant finding the other molecules it binds to, learning how it affects their behavior, and discovering how particular combinations change the activity of genes and other molecules.

Beta-catenin originally comes from the mother and is present throughout the egg. Normally it is broken down by another molecule called GSK-3, which keeps it from activating genes. One way to allow beta-catenin to do its job is to block GSK-3. This is accomplished by the protein "disheveled," which gets shuffled to one side of the embryo after fertilization. There, it locks up GSK-3, leaving beta-catenin intact. It accumulates and docks onto another transcription factor called Tcf3. The combination activates several other genes that lead to the formation of the mesoderm.

FROM THE ORGANIZER TO THE NOTOCHORD

Figuring out the complicated relationships between beta-catenin and other molecules has taken decades of work by many labs and is a good example of the way molecular biologists have tried to understand gastrulation and other developmental processes. Each region of the endoderm, mesoderm, and ectoderm produces a different set of proteins that function as signals to neighboring tissue as it grows and slides by. The information is interpreted according to other molecules present in these cells.

The cells of the organizer move underneath the upper layer, the ectoderm, to begin creating the mesoderm on the back (dorsal) side of the organism. The result is a clear signal to some regions of the mesoderm and the ectoderm that they are at the back. A new migration begins: Dorsal cells begin to move inward. The first cells to do so will begin to form the head. Those that come later will define the tail and the lower part of the body.

Working at the University of California at Los Angeles in 1993, Christof Niehrs and Eddy de Robertis showed that signals from the Nieuwkoop organizer triggered cells in the mesoderm to produce the goosecoid protein. The scientists injected this molecule into various locations in the early frog embryo. It developed a second dorsal side and a second front-to-back axis that ran in a different direction than the original. Sometimes the embryos developed a second nervous system, including an extra head and brain.

The effect was clear, but the identity of the signal was still a mystery. William Smith and Richard Harland at the University of California at Berkeley began closing in on an answer. Their lab exposed frog eggs to ultraviolet radiation, causing random mutations in genes. They examined embryos until they found cases where the dorsal-ventral axis did not form, which showed that they had succeeded in damaging the gene for the signal.

This is an elegant type of experiment that at first might seem backward—they had destroyed what they were looking for. In fact, it gave them just what was needed: a model system that could be used in the search. Smith began injecting molecules into the damaged eggs and watched to see what happened. When he injected the noggin protein, known to be found in the organizer, he "rescued" the signal; cells now began building a proper embryo again. Depending on the amount of noggin that was injected, the embryos developed in different ways. High amounts turned too much of the tissue into dorsal structures, creating embryos that were little more than heads. This raised a new question: Under normal circumstances, what limited the amount of noggin?

Over the next few years scientists discovered chordin, follistatin, and several other proteins secreted by the organizer that also played an important role in creating dorsal tissue. They also found a molecule that seemed to be able to tune down such signals: Bmp4 (short for "bone morphogenesis protein 4"). By docking onto the surface of cells in the mesoderm, Bmp4 induced the development of the skin and ventral structures, such as muscle, the kidneys, and red blood cells. At the same time it blocked organizer signals that would otherwise transform them into nervous system tissues. In parts of the ectoderm where there is no Bmp4 brake, the protein neurogenin launches the development of neural tissue. In 1996 Tewis Bouwmeester and other members of de Robertis's lab began homing in on another protein that was crucial to the development of the head: cerebrus, which triggers the formation of eyes and the nasal passages.

Combinations of these signals lay down the directional system in the body, forming the major axes: anterior-posterior and dorsal-ventral. A region of cells in the mesoderm induces changes in the layer above, preventing some ectoderm cells from joining their neighbors and becoming skin. Instead, they form a tube at the back of the organism that runs alongside the primitive streak. This tube, the notochord, grows longer as the organism stretches in the vertical direction. Its midsections will eventually form the spine; the top will become the head and brain.

SUMMARY

The first stages of differentiation in humans and nearly every other animal species occur when the embryo's blastomeres—its earliest, unspecialized cells—migrate to create three layers. The ectoderm, mesoderm, and endoderm express different genes and communicate with each other via signaling molecules. As they do so, they first define the position of the back of the embryo. In many species the back-to-front axis is already defined by the position of molecules implanted in the egg cell by the mother; in mammals it occurs because cells launch their own genetic programs.

A group of cells called the "organizer" begins broadcasting signals that tell neighboring cells to form more specialized regions in the three layers. The action of these signals has to be carefully limited, otherwise too many cells will specialize in one way, overriding signals that are needed to define other tissues.

Recent work has identified some of the key signaling molecules involved in creating the different tissues. Most of these are transcription factors, powerful proteins that often activate many different genes.

The ectoderm, mesoderm, and endoderm are found in nearly all animals alive today, meaning that they arose in some of the earliest multicellular creatures that lived on Earth. They are crucial to the formation of organs, as will be seen in the next chapter.

3

The Origins of the Major Organs

At the end of gastrulation, discussed in the previous chapter, a human embryo has three layers of tissue and the beginnings of some of the major structures of the body. During the next phase of development, *neurulation,* the layers communicate with one another and shape themselves into the intricate, origami-like folds of the major organs. Tissues separate and stretch and bend. Cells die to create gaps such as those that separate the fingers or holes such as the hollow spaces within bones that will be filled with marrow. It is impossible here to do more than introduce just a few important organs, but other organs form according to the same basic principles. The aim of this chapter is to give a taste of how developmental biologists study the very complex rearrangements of cells that create tissues.

FROM THE NOTOCHORD TO THE NERVOUS SYSTEM

The brain, the spine, and the body's nerves begin in the outer layer of the embryo, the ectoderm. There, on the surface, cells divide at a rapid pace, and suddenly a group of them pushes inside and begins a migration under the surface. They create a new cavity called the *archenteron* by pushing around the other cells of the interior until the earlier cavity, the blastocoel, disappears. The archentron becomes the major opening in the interior of the embryo and will eventually form the gut cavity.

These are not the first cells to move inside, however. Cells that migrated earlier have received instructions that turn them into the middle layer, the mesoderm. Chapter 2 described how they form a thick band called the notochord, which stands vertically between the very back and the gut cavity as a sort of ladder. This creates the anterior-posterior axis of the organism and helps keep the body organized as the embryo grows. In later stages, that function is taken over by the spine, and the notochord will develop into the nervous system. The notochord arose early in animal evolution and served as the main structural support in early types of fish that lacked a bony skeleton and a spine.

The notochord communicates with tissue behind it to build a thick zone of cells called the neural plate. Cell division makes it expand and curl. The edges of the plate begin to fold, creating the neural groove. As growth and folding continues, hinges form on the sides and are pushed together. Molecules on their surfaces recognize one another and signal the edges to fuse together, creating the pipelike neural tube. In 2007 Patricia Ybot-Gonzalez and Andrew Copp of the University College, London, were looking for molecules that controlled this process and found some familiar culprits. Folding is stimulated by noggin, the signal protein that helps "organizer" cells at the back set up the embryo's dorsal-ventral axis (discussed in chapter 2). Additionally, they found that a Bmp protein (also introduced in the preceding chapter) holds up the process of bending. The combination of these two factors determines when and where the folds are made.

The tube becomes fully closed when the human embryo is about 20 days old. Behind it there is still a crease, a sort of footprint left when the ectoderm cells began their migration, but now new cells grow in from the sides to close it. Once the tube is finished, the top bulges and begins building several regions of the brain. The rest will become the spine.

All vertebrate animals have a neural tube, but there are differences between species in the way it forms. In chicks it begins to close at one end and continues to the other like a zipper. In mammals the edges grow together and fuse at many points up and down the line, like buttoning a tight shirt. If the neural tube fails to close properly, the consequences are usually disastrous. If the top does not close, for example, the child will be born without a forebrain. A failure of closure at the bottom results

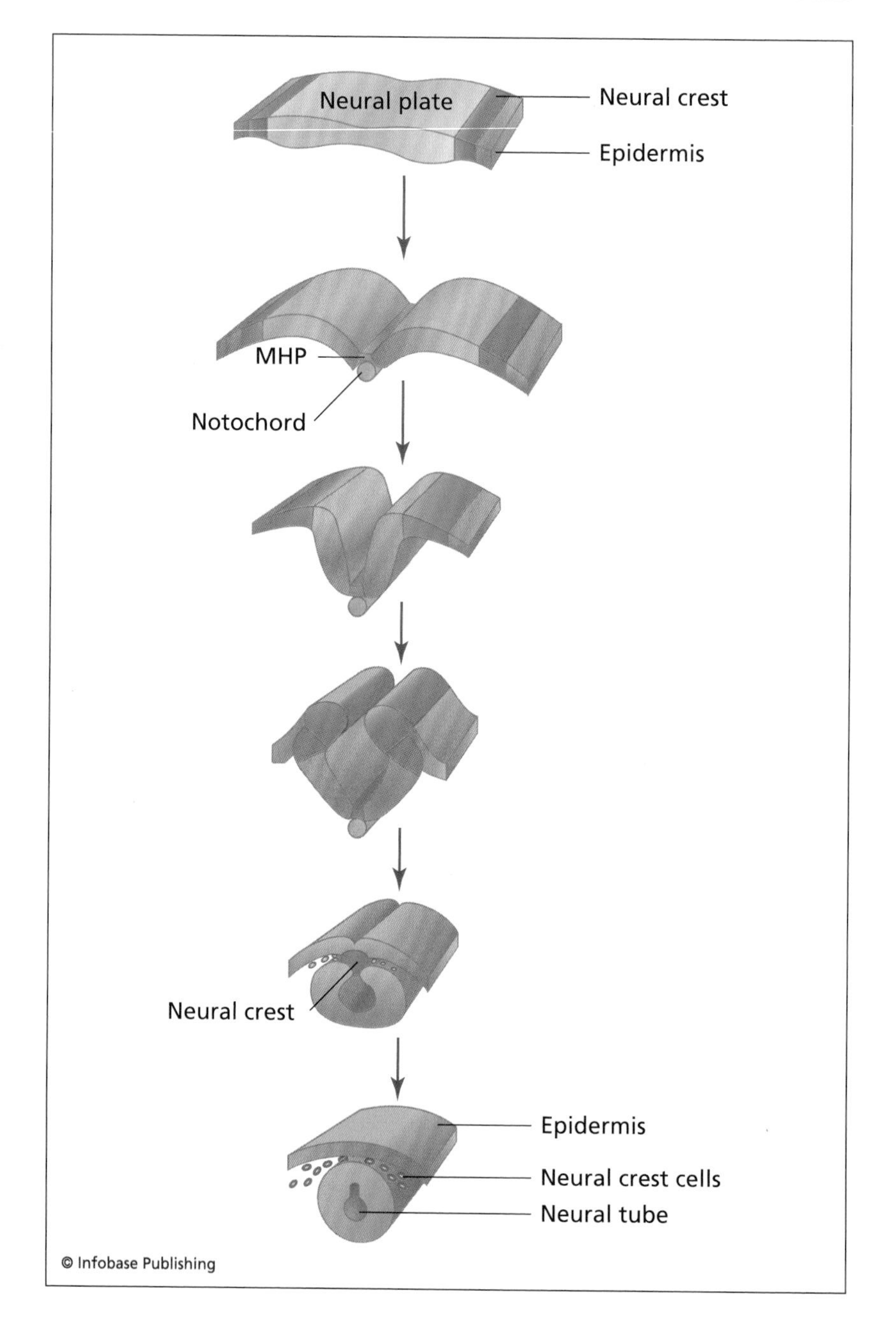

in a condition called "spina bifida." Some of the latter cases are mild, involving very small openings in the spinal column that do not allow the nerves inside—the spinal cord—to protrude. A person may not know there is a problem until adulthood. In more serious cases there are large openings that are obvious at birth or even earlier. The spinal cord emerges, and the result is severe health problems that range from paralysis to accumulations of fluid in the brain.

Spina bifida is probably not caused by flaws in a specific molecule, although families with mutations in PCYT1A, VANGL1, and a few other genes seem to have a higher risk of bearing children with the condition. Other factors have an important influence on the problem. Three decades ago Julian Little and J. Mark Elwood of Aberdeen University in the United Kingdom collected evidence to suggest that children conceived in the winter have a higher likelihood of developing spina bifida than those conceived in other seasons. They also found that the first children a mother conceives seem to have a lower risk than later ones, although this does not seem to be connected to the age of the mother. Little and Elwood also showed that children of mothers from lower socioeconomic groups are more prone to neural tube defects (NTDs). Overall, NTDs appear in about one in 1,000 births, making them the most common type of birth abnormality.

In the 1970s and 1980s some of the connections between spina bifida and diet started to become clear in work headed by Fiona Stanley, an Australian epidemiologist. Stanley carried out careful studies of the health problems of Aboriginal children versus their Caucasian counterparts, looking at diet and social factors as potential contributors to the development of disease. One of her many discoveries was a connection between the lack of vitamin B_9 (folic acid) in a mother's diet and the likelihood that her children will suffer spina bifida. If mothers take folic acid before and during pregnancy, spina bifida becomes

(opposite) The patterns of growth, folding, and pinching that produce the notochord, neural tube, and neural crest cells

much less likely or less severe. Since then the findings have been confirmed several times. This prompted the British government to consider requiring bakers to add the vitamin to bread; it has also led to changes in dietary recommendations for young women in the United States. Stanley's work and the resulting programs have markedly reduced the occurrence of this developmental defect.

Folic acid has been known to play a role in pregnancy since 1931, when Dr. Lucy Wills (1888–1964) discovered that this substance was necessary to keep pregnant mothers from becoming anemic through a loss of red blood cells. In the meantime, scientists have learned that the vitamin is crucial for the type of rapid cell division that takes place in early embryos. In 2007 Kotaro Kaneko and Melvin DePamphilis of the National Institutes of Health in Bethesda, Maryland, showed that vitamin B probably has its effects because of its influence on a molecule called TEAD2. This protein is produced by the mother and enters the embryo, where it activates genes. Mice whose mothers lacked the protein had a much higher risk of being born with NTDs. The risk dropped again—but was not completely eliminated—when the mothers were given vitamin D.

The development of another structure called the *neural crest* accompanies the formation of the neural tube. It arises in a narrow gap behind the neural tube and the skin that closes to cover it. The neural crest only exists for a short time but is extremely important. It has several regions that produce a wide variety of cells that migrate throughout the body. Some of these become the bone, cartilage, and neurons of the face, ear, and jaw. Others wrap themselves around blood vessels, establishing a connection between the growing circulatory and nervous systems. These will play an important role in regulating blood pressure and other aspects of circulation. Yet another type of cell moves to the gut, where it manages involuntary muscle contractions that transport food through the intestines. Finally, a fourth region of the neural crest produces cells that migrate to the heart, where they help connect large arteries to muscle tissue.

SECONDARY NEURULATION AND THE EARLY BRAIN

If everything goes normally and the neural tube closes, the second phase of neurulation begins. At first, the tube is a solid column of dense cells, but by the end of this process it will be hollow. This happens because the cells in the interior draw together, separating from their neighbors and leaving gaps. Soon the gaps join to create one hollow area within the tube, which stretches from the top of the organism to the tail. Initially the tube is long and straight, but parts in the uppermost region begin to bulge to form three sacs, or vesicles. These will develop into the major parts of the brain, the sac at the top becoming the forebrain, followed by the midbrain and hindbrain. The lower parts of the tube become the spinal cord and spinal column.

These first stages of brain development are very similar in most species of vertebrates. Unique features develop later, as cells in the walls of the tube differentiate and migrate. Humans have evolved a unique forebrain, for example, that is responsible for key memory processes and many of the functions related to human intelligence. But long before it arises, cells at the lower end of the first bulge form the retina—the nerves in the eye—and brain circuitry for vision and hearing. They also develop into the hypothalamus, a center that controls sleep, breathing, and other body functions.

The midbrain is the most basic and ancient part of the brain, found in animals ranging from insects to humans. It belongs to the brain stem, which is the link to the spinal column, and is crucial in controlling the body's muscles and nerves. One region of the midbrain, the substantia nigra, produces a substance called *dopamine,* which is a special version of one of the amino acids (the building blocks of proteins). In nearly all known animal species, dopamine acts as a *neurotransmitter,* a substance that brain cells release and absorb to communicate with one another.

The third main sac of the neural tube, which will become the hindbrain, develops smaller bulges. These specialize into brain structures: the cerebellum, pons, and medulla. The cerebellum is the main coordinator of muscle activity in tasks such as staying balanced while walking or running. The pons connects it to the cerebrum, giving people conscious control over their mus-

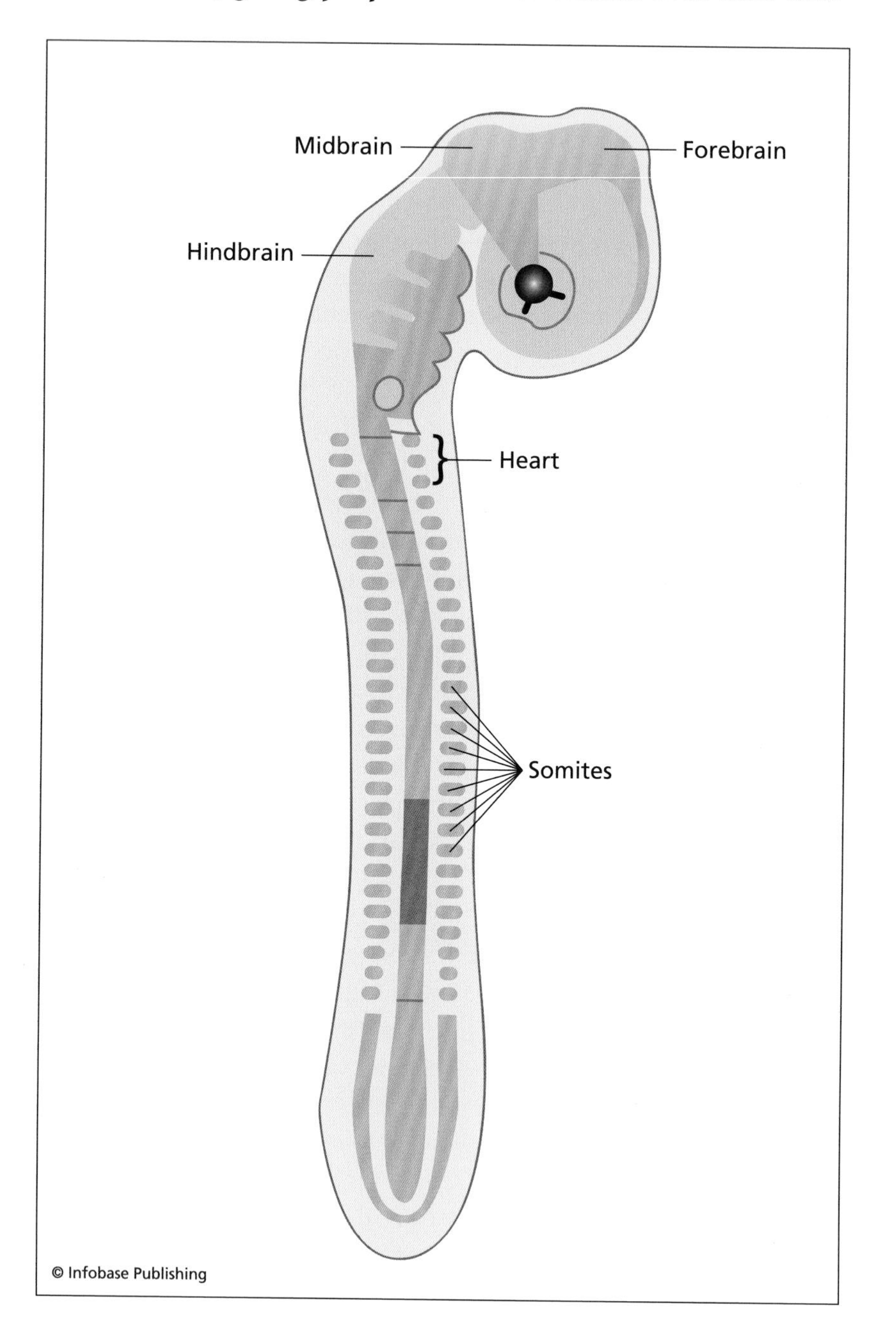

cles and bodies. The job of the medulla is to pass information such as pain between the spinal cord and the brain. It is also responsible for reflexes and involuntary responses such as blood pressure, breathing, and pulse rate.

The neural tube specializes not only in the top-to-bottom direction but also along the dorsal-ventral axis. Once again, this process is guided by different signals: Bmp4 from the back and a molecule called Sonic hedgehog (Shh), transmitted from cells in the notochord, to the front. (Each of these molecules is a powerful developmental signal that helps shape several other organs as well.) Eventually the back parts of the brain take on the job of managing most input from the senses, while the front is responsible for controlling muscles.

The brain and many other organs are extremely sensitive to alcohol and drugs throughout the development of the embryo. Ethanol is the ingredient in alcoholic drinks that causes intoxication; for embryos it is poison. The connection between alcohol and developmental defects had become clear by the early 1970s. Between 1954 and 1957 French medical student Jacqueline Rouquette studied 100 cases of children with developmental abnormalities who had been born to alcoholic parents. In her thesis she concluded that it was especially dangerous for mothers to drink during pregnancy. Ten years later another French doctor, Paul Lemoine of Nantes, carried out a similar study in which he described the condition now known as fetal alcohol syndrome, or FAS.

The name was coined a few years later by David Smith at the University of Washington Medical School in Seattle. His team, which included Christy Ulleland, Kenneth Jones, and Ann Streissguth, independently came to the same conclusions as Lemoine in a major international study published in 1973. Researchers across the world were discovering cases of children

(opposite) Secondary neurulation produces a closed tube with new cells migrating and growing to form three saclike bulges. These will develop into the forebrain, midbrain, and hindbrain. Ball-like clusters of "somites" form alongside the tube; each somite is destined to produce specific tissues.

with malformations in the bones and structure of the face and limbs, along with other symptoms such as slow body development, mental retardation, and hyperactivity. It looked like an epidemic or a genetic disease, but the children came from every country and ethnic group, and there was no connection other than the fact that their mothers were alcoholics.

Today FAS affects more than one of every 750 children born in the United States and is recognized as one of the most common causes of mental retardation. Experiments in mice have shown that ethanol disrupts the migration of neural crest cells and causes the death of millions of neurons. During certain phases of pregnancy the ethanol in a single drink, passing into the mother's body and blood and into the fetus, is enough to disrupt proper developmental processes.

THE DEVELOPMENT OF UNIQUE FEATURES OF THE HUMAN BRAIN

Most of the developmental processes of the brain discussed so far function the same way in mammals and other vertebrates, but early in development some unique things happen in the human brain that will play a role later in its unique features and abilities. First, there is a migration of cells from the uppermost region of the neural tube downward. They become part of the thalamus, a major brain coordination center that is much larger in humans than in closely related species of primates. It is extensively "wired" to the cortex, an outer brain layer that develops from the population of cells that remain at the top of the tube. The thalamus is thought to gather input and "translate" it for the cortex. Scientists also believe that it plays an important role in consciousness. Damage to this tissue usually leads to a coma.

An important characteristic of the human brain is its large size compared to the rest of the body—the ratio is greater than that of any other animal, even dolphins and whales. Bruce Lahn of the University of Chicago says that much of this can be attributed to a sudden evolutionary event that happened about

1.5 million years ago when the brain of a hominid ancestor suddenly tripled in size. Most of the increase occurred in the cortex.

Interestingly, at earlier stages in life there is not that much difference; newborn humans and primates have about the same overall brain-body ratio. But besides structural differences in the makeup of the organ, there are some important "qualitative" differences. For example, other species' brains are almost completely developed at birth. In humans, cells continue to create and connect neurons at an incredible pace. During an infant's first year every cubic centimeter of the cortex experiences tens of thousands of new connections. This flexibility certainly contributes to the amazing amount of socialization and learning displayed by human infants. It also has a cost: Newborn humans are dependent on the care of their parents for much longer than other species. Infant primates usually become alert and mobile almost immediately after birth. It is almost as if, compared to apes, humans are born too early in their development—a point of view proposed in 1941 by the Swiss zoologist Adolf Portmann. From the point of view of the mother, this is crucial because her body would never manage to give birth to an embryo with a head the size of a one-year-old child's.

All of these developments are far in the future for the early neural tube. It is home to generic cells that specialize to become neurons and glia, the main types of cells found in the adult brain. Originally the tube consists of a single layer of cells, but they divide to create more layers. Each duplication produces one stem cell that remains in the region around the tube and a second that specializes and migrates. The earliest of the differentiated cells usually stay nearby; later offspring may travel long distances. Thus the brain forms from the inside out as more and more cells are created and spread into thickening zones, forming three distinct layers around the tube.

The spine keeps this shape throughout life, but cells that become the brain behave in a much more complex way. After creating major structures like the cerebrum and cerebellum, they continue to specialize and migrate. The cerebrum eventually has more than 40 modules that have distinct jobs because their neurons are linked to specific tissues in other parts of the brain and the rest of the body. The principles by which these modules form are similar to the developmental processes that will be seen in all other organs: Cells in one region stimulate the formation of their neighbors through an exchange of signals. New transcription factors are activated that give "suborgans" of the brain their individual characteristics.

The main difference between the development of human and other primate brains is not genes themselves. Chimpanzees have a close relative to nearly every human gene, but they are used in different ways. Until recently the differences were hard to assess, because researchers could only compare the activity of a few genes at a time. Since the completion of the human genome, new types of biotechnology have made it possible to obtain "readouts" of the complete gene activity of cells. In 2003 the laboratory of Carollee Barlow at the Salk Institute in La Jolla, California, compared the behavior of genes in human and chimpanzee brains and came to a surprising conclusion. The main difference lay in 169 genes. Most of these were active in the brains of both species, but humans were using more than 90 percent of them at a much higher rate, producing far more messenger RNA (mRNA) molecules. This was not the case for the heart and liver, where genes were nearly equally productive in the two species.

Geoffrey Woods of the University of Leeds has been looking for unique features of the evolution and development of the human brain by investigating people in whom the process has gone astray. Microcephaly is a condition in which the brain grows to only a fraction of the size found in most people—in the most extreme cases, it is only one-third as large as an "average" brain. Careful studies by Woods and his colleagues revealed that microcephaly seemed to be inherited, which strongly suggested that defects in genes might be responsible. In 2002 the laboratories of Woods and Christopher Walsh at Harvard discovered that mutations in the ASPM and microcephalin genes could cause the condition. In mice ASPM has a crucial role in helping early cells develop into neurons.

Two years later the scientists collaborated with Vladimir Larionov's laboratory at the National Cancer Institute in Bethesda to investigate the evolutionary importance of the ASPM gene. By comparing the makeup of the gene in many different primates, they concluded that it had undergone mutations that probably played an important role in the great increase in size of human brains.

Other genes seem to be linked to specific human abilities. A unique form of a gene called FoxP2 plays a crucial role in language. In 2001 the laboratory of Anthony Monaco at the Wellcome Trust Center for Human Genetics in Oxford, United Kingdom, showed that people who inherit particular mutations in this gene have great difficulty speaking and forming grammatical sentences. More recent studies by Svante Pääbo's laboratory at the Max Planck Institute in Leipzig, Germany, show that the gene has undergone a high number of changes during recent evolution.

HOX GENES: A UNIVERSAL SCAFFOLD FOR THE HEAD-TO-TAIL AXIS

A common type of children's book lets readers create strange creatures by flipping parts of the page to mix different parts of animal bodies. The pages are split so that a bird's head, for example, can be matched to a dog's torso and human legs; by flipping part of the page, a child can change the head to that of a tiger and the feet to those of a duck. Bodies come in segments that everyone recognizes, and the reason can be traced back to a set of molecules known as the *homeotic genes* in insects. The name comes from the term *homeosis.* It was coined by William Bateson (1861–1926), a geneticist of the early 20th century, to describe strange mutants with rearranged body parts. In the meantime, similar molecules have been found in a huge variety of animals, and today they are known as the "Hox" genes.

In the 1970s geneticist Edward Lewis (1918–2004) of the California Institute of Technology (Caltech) discovered that certain genes in the fly have a tremendous impact on the construction of parts of the body's architecture. By cross-breeding particular strains of flies, he was able to obtain insects that lacked an entire group of molecules called the "bithorax genes" needed to define parts of the back and lower body. Lewis began adding various combinations of genes that restored some of the segments. This led him to a surprising discovery: The bithorax genes were arranged on a chromosome in the same order as the positions of the body parts they controlled. It was a bit like entering an architect's office and finding plans for the top floor of a building on the top shelf, the basement on the lowest, and all the intervening floors arranged in order in between. No one had suspected that the physical location of genes might have anything to do with the structure of the body, but, indeed, the first Hox gene on the chromosome was responsible for the most anterior structures, the next gene controlled the segment that immediately followed it, and this continued all the way to the tail. Originally eight genes were found that worked together in this way.

Until the 1970s Lewis and other scientists had mostly concentrated on how genes affect the body plans of adult flies and other organisms. But, whereas adult features arise over the course of development, studying bithorax genes might provide clues as to how this occurs. Lewis began working backward to investigate the effects of the genes in embryos. He discovered that fruit fly larvae are clearly divided into beltlike segments. Each develops into particular body structures in the adult. Nüsslein-Volhard and Wieschaus were also studying how mutations in genes affected the development of these segments. Their work brought the fields of genetics and development together and earned the three scientists the 1995 Nobel Prize in physiology or medicine.

Walter Gehring and his colleagues at the University of Basel, Switzerland, made the next important step in understanding the contributions of Hox genes to development. First, the laboratory discovered that each of the homeotic genes contained a nearly identical sequence that they named the *homeobox*. Matthew Scott and Amy Weiner, working with Thomas Kaufmann at Indiana University, were coming to the same conclusion independently. Allen Laughon, a professor at the University of Wisconsin, was first to realize that the sequence was very similar to codes that allowed proteins to bind to DNA. He figured out that homeobox genes encoded transcription factors, proteins that had their effects by docking onto other genes and changing their patterns of activation. So, activating each gene provided a key that could unlock the next one and the next set of instructions in the developmental tool box.

Portrait of Dr. Walter Gehring. Gehring's laboratory in Basel played a crucial role in explaining the evolution and functions of Hox genes. *(Walter Gehring)*

Bill McGinnis and other members of Gehring's group wondered whether these molecules were unique to flies, so they began looking for similar genes in other organisms. They quickly turned up case after case in humans, cows, mice, and frogs. Nearly all of the molecules, the Hox genes, are involved in setting up body patterns in these diverse species. A subset of Hox genes called the "Hox cluster" is responsible for head-to-tail organization. Others have roles to play in creating arms and legs, eyes, and the heart. The discovery was similar to finding that architects everywhere arranged their shelves in the same way, no matter what type of building they were working on.

The genes in the Hox cluster are like lines of code in a computer program that have to be carried out in a particular order at a particular time. If one line of code is missing, a segment of the body will be missing or malformed, and the instructions that follow may be misinterpreted. Many genes that work together to establish parts of the body are scattered throughout

The genetic programs that build antennae and limbs in flies are almost identical. Mutations in a single gene called antennapedia cause legs to be built in place of antennae on the fly head. *(F. Rudolph Turner)*

the chromosomes, but the Hox genes have stayed neighbors in most organisms. This is unusual because over the course of evolution random rearrangements of chromosomal material usually separate neighboring genes. The information carried in Hox gene clusters is so important that losing a part of the message or changing the order of its words is usually fatal to an organism and is therefore not passed along to any new offspring.

The genes in the clusters are used all along the axis of the body, from the beginning of the hindbrain down to the tail. They are activated in the neural tube, the neural crest, the mesoderm, and the skin. Their precise functions have been determined by several types of experiments. One method involves "knocking out" single or multiple genes (removing them from an embryo using genetic engineering methods); another strategy is to make them become active in the wrong places. More insights have come from comparing the development of species that use them in different ways.

In flies Hox genes are responsible for structures such as antennae, legs, wings, and parts of the mouth. Sometimes very small differences in the behavior of these genes produce significant differences in body structures. If an organism acquires extra copies of one of them, the result may be extra wings, body segments, or other structures. The Hox gene Pax6 is a master controller for the development of the eye; if it is active in the wrong tissues, extra eyes may develop in the wrong places. A gene called antennapedia determines where legs develop in the fly. If a mutation disables parts of the gene, antennae develop in place of the fly's legs. Mutations that increase its activity cause legs to develop on the head in place of antennae. Such transformations are also interesting from the point of view of evolution; once a structure is in place, only one mutation (or a few) may be needed to transform it into something else with a different but very useful structure and function. Other variations are useless, a nuisance, or dangerous.

In 1991 Osamu Chisaka and Mario Capecchi of the University of Utah began systematically studying the Hox genes in mice. Removing each gene resulted in the loss of a different segment: glands in the hindbrain, parts of the spine, or other structures along the back. Capecchi and others went on to show that cells were properly migrating into the tissues from the neural crest, the first step in forming these structures, but somehow the cells had lost the instructions telling them what they should become. In some cases another Hox gene could fill in the missing instructions. This revealed that Hox genes and partner molecules needed to work together to build a particular body structure.

Vitamin A (retinoic acid) is required to activate some Hox genes, particularly in the hindbrain, and has a strong effect on many others. This has allowed scientists to turn it into a powerful tool to control and investigate Hox genes in laboratory

The order and location of body segments of insects such as this centipede are determined by the sequential activation of Hox genes. *(C. L. Hughes, Salk Institute)*

animals. Adding retinoic acid to organisms allowed scientists to observe what happened when Hox genes were too active or active in the wrong places. In fish this caused Hox genes to become active too far forward, leading to deformities of the face.

Human embryos can also be influenced by retinoic acid. It is crucial to the development of the retina, and children who do not get enough of it from their diet may become blind or suffer other health problems. On the other hand, overdoses at the wrong times of a woman's pregnancy can cause problems. It adversely affects the blastocyst and can have negative effects on the development of the heart and other tissues. A common medicine for acne was found to contain the substance; women who used it during pregnancy exposed their fetuses to high levels. Many of their children suffered developmental defects of the face and some parts of the brain, echoing effects seen in fish and other model organisms.

FROM LEFT TO RIGHT: SYMMETRIES AND ASYMMETRIES

Retinoic acid is also involved in making sure that the left and right sides of the body are symmetrical—and sometimes that they are not. Most of a person's left side is a mirror image of the right, but many of the internal organs lie on one side or the other. In 2005 geneticist Pascal Dollé's laboratory at the Institute of Genetics and Molecular and Cell Biology in Strasbourg, France, studied the development of mice whose bodies were unable to produce the substance. They discovered that the timing and formation of structures on the two sides became desynchronized; the right side was slower.

The embryo is symmetrical from left to right until the heart begins to form from a tubelike structure. Then cells on either side of the midline begin to produce different molecules; the result is that in most people, the lower half of the tube bends to the right. Eventually this leads to a heart in the left half of the chest cavity and a liver in the right, but a small percentage of people have a reversed body plan called "situs inversus viscerum." In 1959 Katharine Hummel and D. B. Chapman of the Jackson Laboratories in Maine discovered that mutations in the situs inversus viscerum gene could scramble the locations of organs in mice. The mutation had

While the hearts of most humans develop in the left side of the chest, in rare cases a mutation reverses the position of the heart and other major organs, as seen in the X-ray below. *(Department of Anaesthesia and Intensive Care, Chinese University of Hong Kong)*

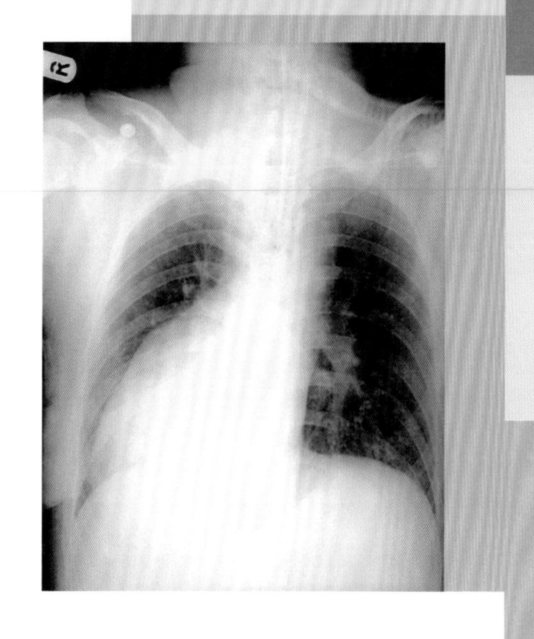

a different effect on each organ, putting them randomly on either side. This meant that the connections between them could not be built, and the embryo usually died. Another gene called "inversion of embryonic turning" switches everything in a mirrorlike way; all the organs commonly found on the left side are shifted to the right, and vice-versa. The animals are put together backward but are otherwise healthy. In humans we now know that mutations in left-right (LR) pathway genes or those involved in development of crucial LR structures such as the midline, node, and nodal cilia are found in approximately one in 10,000 live births.

Interestingly, some humans whose left and right sides are reversed also suffer from Kartagener syndrome, a problem with cells in the respiratory tract, the fallopian tubes, and other tissues whose cells have hairlike cilia. Normally these microscopic structures need to beat—like the strokes of an oar—to move fluid along in a consistent direction. That does not happen in people with the syndrome, and the connection between these two unusual features turned out to be a clue about why the heart and other organs have asymmetrical positions.

The discovery came in 1999 from Nobutaka Hirokawa's laboratory at the University of Tokyo in Japan. The researchers showed that cilia on cells in the "embryonic node"—the equivalent of the Spemann organizer in mammals—carry out a beating motion that pushes fluid from right to left. In 2005 they showed that this carries along retinoic acid and other molecules that provide cells with developmental signals. The details are not yet clear, but cells often respond to different concentrations of developmental signals in different ways. This means that different genes will be activated in cells upstream and downstream of the "nodal flow," eventually creating differences between the two sides of the body. As their cells take on separate developmental paths, the midline of the body, which runs along the ventral side of the neural tube, insulates the two sides from each other.

THE HEART

A few species of animals can survive without a heart and circulatory system. Some species of worm manage this by being so flat that all of their cells come in contact with the environment or are just a few layers away. This allows them to absorb nutrients directly from their surroundings. Sponges also get along without a heart because they are perforated by a huge number of pores that deliver seawater to within a millimeter of every cell. But most organisms have had to solve the problem of providing their cells with food and oxygen and disposing of wastes through the construction of an intricate system of internal canals and a pump that keeps fluid moving along through them.

The problem of supply arises very early in animal development once the organism has become thicker than about 200 microns, about four times the width of a human hair. At that point all of its cells no longer have equal access to the environment. The solution is that one of the layers created during gastrulation—the mesoderm—is transformed into major parts of the circulatory, respiratory, and digestive systems. As usual, the mesoderm does not manage this alone; it relies on molecular signals from neighboring tissues. It also requires the development of specialized cell types, which will be covered in the next chapter.

Without a heart, blood would not move through this complex system; it would stagnate and could not be loaded with fresh oxygen and other nutrients. It has to be pumped through, and larger animals require more pressure to move it. Evolution produced various solutions, ranging from simple muscle-lined tubes that squeeze blood through the body to the complex, four-chambered organs of humans and other mammals. Comparisons between different types of hearts reveal some of the steps in these transformations. Each type and developmental stage is a solution to a complex engineering problem. The heart has to function very early in life to get nutrients to cells. Because the embryo grows so quickly, it undergoes several major transformations. The circulatory system is a massive work in progress. New routes are continuously built to supply an ever-growing population of cells. And then there is the problem of birth, which ejects an embryo that has been living in water into the air. The entire circulatory system is transformed—within minutes—from a closed "circuit" that feeds off the mother to one able to cope with oxygen and food.

The heart's origins can be traced back to events between the second and third weeks of the human embryo's life, as the neural tube takes shape. At that point the embryo is stretching to become a long, bumpy oval. Seen in cross-section, the ectoderm is a bit like one bag containing another bag (the mesoderm); this tissue, in turn, holds the endoderm and the hollow space that will become the gut.

The mesoderm thickens a bit and opens in the middle, like a slowly inflating air mattress. Two of these mats develop on the sides of the notochord (see image on page 70). The space in between will soon become a tube, formed from mesoderm tissue growing inward from right and left. Before that occurs, a bit of underlying tissue from the endoderm is squeezed up through the space between the mesoderm and the notochord. To go back to the air mattress analogy, if the two mattresses are lying on a carpet (the endoderm) and are pushed together, the carpet might wrinkle between them, making a bulge. In the organism this bit of tissue will be pinched off. The top of the bulge will develop into the foregut, which will later become the respiratory system, the stomach, and several other organs.

Now the two mats need to grow together. The mesoderm pushes in from the two sides. It does not completely fill the gap; a hollow space is left in the middle, creating the tube in between. This leaves the organism with three tubes that grow vertically through the body. Seen from the top, this has produced the neural tube on the dorsal side, which will develop into the spinal cord; the foregut in the middle; and the heart on the ventral side. The movements of endoderm have given the heart tube an inner lining (the endocardium) and an outer one (the myocardium).

As its cells differentiate into muscle, they begin to pulse. Single cells have the unusual ability to contract by themselves, and they will continue pulsing rhythmically even if removed from an animal embryo and placed in a test tube. In the animal they find a common rhythm by communicating with each other by releasing and absorbing two substances, calcium and sodium, in waves. These are the embryo's first heartbeats, and they take place long before the heart has developed any familiar structures. It does not yet pump blood, however; that will start about a week later.

The heart of an adult human is a sac of thick muscle with four inner chambers separated by valves. It is connected to the circulatory system by large arteries and veins. Building it from a simple tube involves bending and separating it with valves and septal walls it as it grows. At first, it bulges in some places and is tightened in others, resembling the relatively simple hearts of fish. Later, two major compartments form, like amphibian hearts. Snakes and turtles have three-chambered hearts. In some types of aphids the organ is much more complex, with eight. (At least two cases have been documented of humans with five-chambered hearts, probably arising through mutations.)

At the beginning of heart formation the embryo has already developed differences between the anterior and posterior ends and the dorsal and ventral sides. In most people the heart sits slightly to the left of center. It begins in the middle but by the eighth week has moved to the side. The spleen, stomach, gut, lungs, and other organs also develop in an asymmetrical way. Their positions are determined by the behavior of cells in the node very early in development and the activity of specific genes.

Imagine placing a long, skinny balloon in a small box and blowing it up: It would soon run out of room and bend. The same thing happens with the heart; the cardiac tube quickly outgrows its space and twists into a loop. Cells in different areas begin turning out specialized proteins. In mice a molecule called Hand1 appears in the left ventricle, while Hand2 is produced only on the right side. In 1998 Deepak Srivastava, a geneticist now at the University of California at San Francisco, showed that if Hand2 is deleted from mice, the right ventricle is lost. The left ventricle remains, but it produces only Hand1. When this happens, the heart tube stops growing. The reason seems to be that Hand proteins interpret asymmetric positional information in the developing heart and participate in development of segments of the heart tube, which give rise to specific chambers.

In 2007 Srivastava and his colleagues found that another type of molecule, called a *microRNA,* also had a huge impact on heart development. Besides their small size, microRNAs are un-

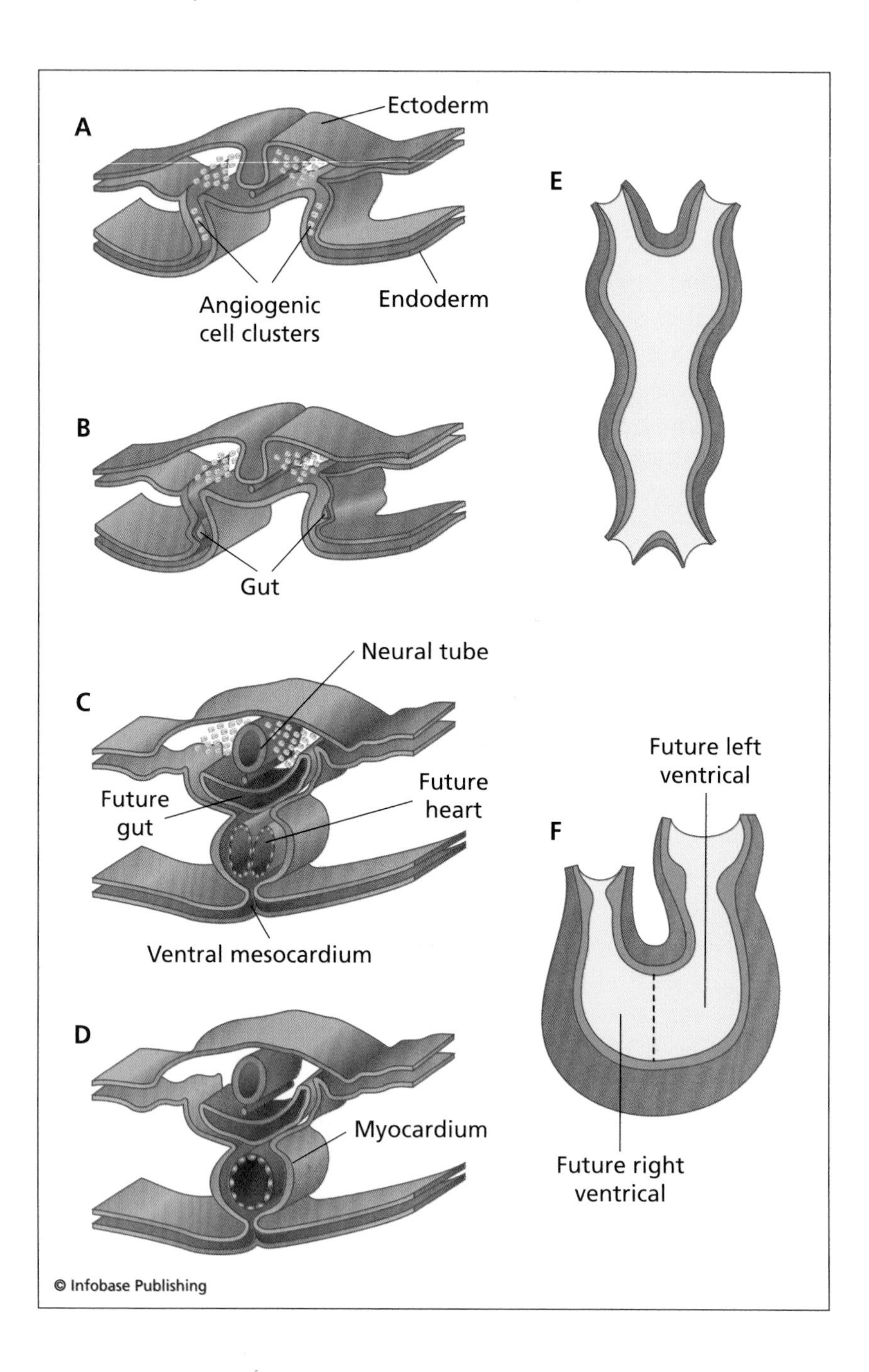

like other RNAs produced by cells because they are not used to make proteins. Instead, they dock onto other RNAs and prevent them from doing so.

MicroRNAs were discovered accidentally in the late 1980s when scientists at a biotech company in California were using genetic engineering to try to give pale purple petunias a much more intense color. Rich Jorgensen's lab inserted an extra copy of the gene responsible for purple pigment, expecting that more genes would produce more of the protein. Instead, the result was a completely white flower without any pigment at all. A look into the cells showed that they were actually using both genes to produce RNA molecules but that the latter were somehow interfering with one another. Labs across the world were discovering the same phenomenon in other plants. The first microRNAs were regarded as oddities; since then thousands have been found in the human genome, and they are known to play a crucial role in shutting down particular genes throughout the body.

Srivastava discovered that a microRNA called miR-1-2 was active only in the heart. When his laboratory removed it from mice, the result was disastrous for development. There was no longer any control over the rate at which cells divided, so the heart's walls became too thick. It beat more slowly than normal because cells communicated inefficiently with their neighbors. Furthermore, the heart chambers were full of holes, the most common inherited heart defect in humans. Srivastava's study suggests that human microRNAs might also be involved in heart defects. MiR-1-2 is known to bind to the RNA of Hand2 and block the production of the protein.

In the meantime, scientists have learned to use this mutual "interference" that takes place between RNAs to block the production of specific proteins. They create artificial *small interfering*

(opposite) Heart development. *A–D:* The gut and the heart form below the neural tube (blue) in a complex series of folds and pinching. They are made from mesodermal (pink) and endodermal (yellow) tissue. *E–F:* Once the double-layered tube of the early heart has formed, it begins to bend and fold, creating chambers and substructures.

Developmental Molecules and Heart Disease

Studies of gene mutations have revealed common genes in different organisms that play important roles both in the healthy development of the heart and in the development of diseases. A gene called Tinman in the fruit fly contains the recipe for a transcription factor that is required for the specialization of mesoderm tissue and the formation of the heart. If Tinman is defective, the "dorsal vessel," which is the fly equivalent of a heart, does not form. In 1995 Ian Lyons and Richard Harvey, geneticists at the Walter and Eliza Hall Institute of Medical Research in Victoria, Australia, identified a mouse gene called Nkx2.5 that is related to Tinman. Nkx2.5 is usually active in the heart from its earliest stages through adulthood. Without it the mouse cardiac tube does not form a tube or differentiate. A study of human patients with congenital heart disease has now shown that some of them have inherited a mutation in the gene.

Transcription factors often work together to switch genes on or off, and Nkx2.5 often pairs up with a family of related molecules called Tbx genes, so it is not surprising that mutations in one of these genes, Tbx5, also leads to heart disease. Such mutations have been found in humans with a condition called Holt-Oram syndrome. Here there are numerous problems with heart development, including a failure of the left side to grow normally and valves that are built in the wrong places. Intriguingly,

RNAs (siRNAs) and insert them into cells, where they shut down specific genes. The method cannot be used in many situations and all organisms, but when it works, it is often technically easier than other methods of knocking out a gene. Researchers are

patients with Holt-Oram syndrome often have malformations in their hands as well. The connection in this case was found in 2006 by geneticist Sylvia Evans's lab at the University of California at San Diego. Her study showed that a signaling protein called Bmp works with Tbx not only to create parts of the heart but also to form limbs. In animals that lack one of the molecules, neither cardiac tissue nor the hind legs develop properly.

If it seems strange that such radically different parts of the body depend on the same genes, one only has to think of computer programs. The same commands may be used in programs to deal with texts, images, or music. Bmp and Tbx are lines of code that appear very early in development. Their function seems to be to write "mailing addresses" on cells that are born near the notochord and need to be delivered to the heart and limbs to help build these structures. Without proper addresses, the cells may be delivered to the right destinations—eventually—but they do not know what to become.

The body uses transcription factors and particular signaling systems in many different tissues and contexts. Embryonic development manages to build a huge variety of body structures with a limited set of commands, and the instructions often have similar results in many species. This is what makes animals useful in the investigation of human diseases, and such discoveries will expand researchers' understanding of birth defects and the relationship between genes and disease. These themes will appear many times throughout the rest of this book.

trying to convert it into new forms of therapies to block dangerous molecules produced by mutations. While trials have already been undertaken, there are many technical difficulties to work out before such molecules can be used as drugs.

BUILDING BLOOD VESSELS

In their early stages the arteries and veins that deliver blood through the body are like small country roads that gradually become woven into a superhighway system with the heart at the center. Tissues in the embryo grow so quickly that the body is a mass of construction sites, scrambling to keep up with the need to push routes into expanding tissues. Soon the entire body has become networked in a single system, but building continues as the body grows. By adulthood controls are placed on this system to slow it down; road crews are mostly called in only to deal with emergencies, such as injuries. The exception is tumors, which grow much faster than healthy tissues and have to be supplied with blood. Vessel building (*angiogenesis*) escapes these controls and behaves more like the process in embryos. Without blood, tumors die, so researchers have invested a great deal of effort in unraveling the mechanisms that underlie angiogenesis, hoping that they will lead to powerful strategies to fight cancer.

Building the circulatory system involves engineering problems similar to those faced by a city waterworks. Even if water is sent out at a huge pressure, the laws of physics dictate that pressure will drop as the system of pipes gets longer and more houses are hooked up. In the body this system is immense; if all the blood vessels in a human were stretched end to end, they would stretch about 60,000 miles (100,000 km), and 100 trillion cells have to be supplied. Raising the pressure very high would burst the pipes or send blood rushing by cells so fast that they could never absorb nutrients. It would also be dangerous because blood would spurt quickly from any wound.

Evolution has found an elegant solution through the development of "pipes" of different sizes. The system can handle the pressure because the large vessels branch off into thousands of smaller ones, having an effect much like punching small holes all along a hose. Major arteries and veins move large quantities of blood at a fast rate. They bud off into tiny capillaries that are the major suppliers to cells; the canals are so small that fluid moves through them slowly, giving cells time to absorb what they need. (Ten capillaries lying side by side would be about the width of a human hair.) Their walls are extremely thin, allowing nutrients and oxygen to pass through. When the nutrients have been exhausted, the blood is returned to the heart. The heart pumps this spent blood to the lungs along a route called "pulmonary circulation"; the network that delivers enriched blood to cells is called "systemic circulation."

Building and connecting these types of vessels is an intricate process involving several types of cells. Blood and vessels arise from the same tissue, the lateral plate mesoderm. Both begin as one type of cell called a "hemangioblast." Large numbers of these cells form pools called "blood islands." Those in the center will become *hematopoeitic cells,* a type of stem cell that generates dozens of types of blood cells. Those at the edges of the islands will form themselves into the vessels that contain and transport blood.

A blood vessel is a multilayered tube assembled from several types of cells in a process called "vasculogenesis." Endothelial cells, which will make up the inner lining, arise at the edges of the blood islands. Endothelial cells are a specialized subtype of epithelia, the types of cells that make up the skin and line body cavities. Once they have specialized, they recruit and wrap themselves in smooth muscle cells. In the next stage they link up into a network that routes blood into vessels that supply the growing heart.

Several genes are responsible for these developmental steps. The German developmental biologists Ingo Flamme and Werner Risau, working at the University of Bochum, did an experiment in 1992 showing that a protein, FGF (fibroblast growth factor), was needed for the development of vessels. They isolated undifferentiated mesodermal tissue from bird embryos and grew the cells in the test tube. When they added FGF, blood islands formed. If they allowed the cells to come in contact with each other, tendril-like extensions grew between them—necessary to sew the cells together into vessel linings.

Another protein, VEGF (vascular endothelial growth factor), has an equally important role in creating blood vessels. This molecule is secreted by cells at the edges of the islands and

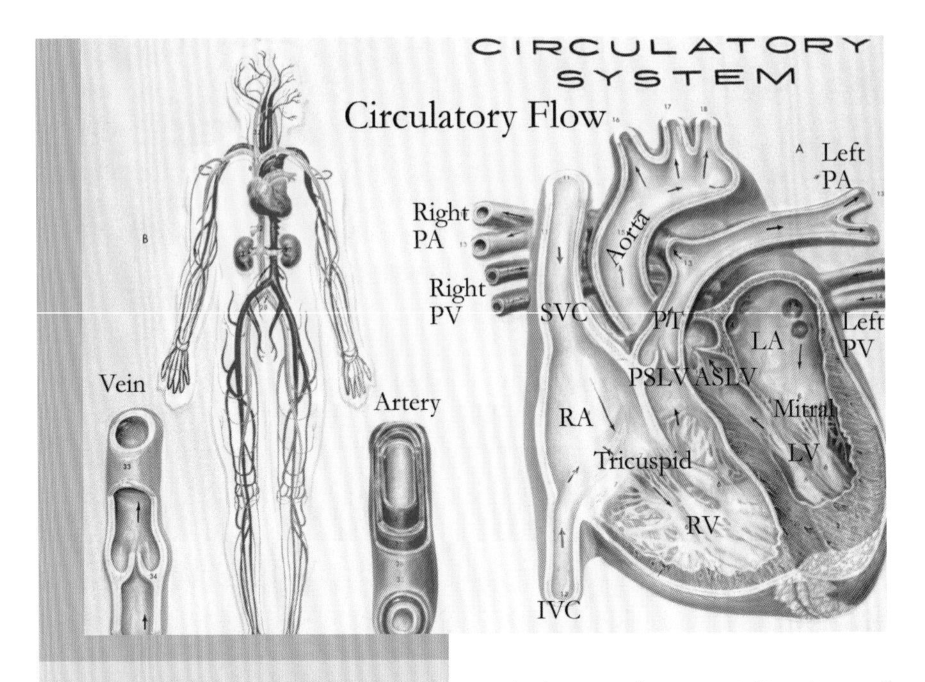

Circulatory system. Left: The system of arteries and veins. Right: A detailed view of the structure of the fully developed heart. *(National Institutes of Health)*

helps in the specialization of the endothelial cells and their linkage into tubes. The molecule is recognized by at least two receptor proteins on the surfaces of developing blood cells. The two receptors have different functions. Guo-Hua Fong and members of Martin Breitman's laboratory at Mount Sinai Hospital in Ontario, Canada, showed that if one of them (Flt1) is missing, the cells specialize into the proper types, but they are unable to link up into vessels. Finishing off the vessels requires the activity of additional proteins and receptors. In 1996 Miikka Vikkula, Laurence Boon, and Bjorn Olsen of Harvard Medical School showed that interfering with angiopoeitin proteins or their receptor, a protein called Tie2, prevents the endothelial cells from recruiting the smooth muscle lining. The protein PDGF (platelet-derived growth factor) and its receptors contribute to this process as well.

VEGF also has an important role to play in angiogenesis, the second major part of the development of the circulation system. Once early blood vessels exist, new ones can be formed by sprouting offshoots that supply new, rapidly growing tissues—like building branches off the main pipes in a city water system to reach new houses. One effect of VEGF is to loosen connections between endothelial cells and disconnect them from the surrounding tissue. In the gaps new endothelial cells grow and shape themselves into vessels. The same molecules that build branches off capillaries are involved in fusing them into larger vessels. This connects them into a unified network, creating veins and arteries. Several additional proteins are necessary to stabilize the connections.

BLOOD VESSELS IN CANCER AND OTHER DISEASES

Tumors grow faster than healthy adult tissue but need just as much blood to survive. The body's normal pace of making new vessels is too slow, and tumors have to speed the process up somehow. One method they use is to secrete VEGF, and drugs that block this molecule are used in the treatment of cancer. In 2007 Darren Browning and his lab at the Medical College of Georgia found another possible point of attack. Browning discovered that tumor cells often shut down a protein called PKG, whose normal function is to keep VEGF in check. Reductions in the amounts of PKG raise the production of VEGF, and this allows the construction of new blood vessels. If scientists can learn to manipulate this pathway only in cancer cells, it might give them a way to starve tumors to death.

The system that controls blood vessel growth is also disturbed in other diseases. If cells begin to turn out too much VEGF, vessels may appear in the wrong places or existing ones may become leaky. In a disease called "wet macular degeneration," a common cause of vision loss for people over the age of 50, too many capillaries expand into the retina. Recently scientists have successfully developed a therapy that uses an siRNA to block cells' production of VEGF.

Tumors rely on other vessel-building mechanisms that are active in embryos but slow down or disappear in adults. Veins and arteries are one-way roads. How does a single vessel know which one to become, and how do the two types become linked? In 1998 Hai Wang and David Anderson of Caltech were able to trace specific blood vessels back to their origins. They found that early vessels that produce the protein ephrin-B2 will become arteries. Future veins make a receptor protein called EphB4 that ephrin can dock onto. (This "key-and-lock" relationship between ephrin and Eph receptors plays a major role in the development of several other types of tissues; for example, it helps wire neurons to each other in the brain, as discussed in the next chapter). The binding of these molecules has several effects. First, it helps connect veins and arteries through "bridges" of capillaries. Second, it shows which capillaries are allowed to fuse together to become much larger arteries and which can become veins.

The discovery of ephrin and Ephs helped explain how cells knew their identities as belonging to one kind of vessel or the other, but it did not explain what caused the types to become different in the first place. Tao Zhong and Mark Fishman of Harvard Medical School tracked the decision back to the "Gridlock" molecule, which is produced in the mesoderm near the notochord. When they lowered amounts of this molecule in zebrafish, vessels became veins rather than arteries. Nathan Lawson and his colleagues at the Massachusetts Medical School discovered similar effects when they interfered with another protein, "Notch," which is crucial to developmental processes throughout the body. This made sense because Notch activity switches on Gridlock.

In 2007, studies by Lawson and his colleague Arndt Siekmann, as well as by several other laboratories, discovered how Notch influenced the sprouting of new blood vessels. Angiogenesis begins with the creation of gaps in the walls of vessels, described above. Tip cells push forward through the gap and extend fingerlike protrusions that act as sensors; the new vessels know where to grow by sensing different concentrations of VEGF. Notch apparently holds this process back by keeping some cells from rushing forward into the tip. If the protein is missing, too many cells push forward, leading to tissues that are overloaded with too many arteries.

Arteries seem to develop before veins, and Lawson and Brant Weinstein of the National Institutes of Health in Bethesda have recently put forward a hypothesis that may explain why. VEGF and the Notch pathway stimulate cells to form arteries. Once these vessels exist, proteins on their surface may call up nearby uncommitted cells that produce the Eph receptor, prompting them to become veins.

In adults Notch does not seem to be involved in creating new blood vessels, but it becomes active again in tumors. This suggests new targets for cancer therapies. Shutting down Notch itself would likely cause more problems than it solved because the molecule has a wide range of other important tasks in other adult tissues. But Notch is activated by other molecules, and it has its effects on the cell after passing signals through many other proteins to genes. Scientists still have not identified all of the genes influenced by the molecule, but Lawson and others are using new techniques such as *DNA chips,* or "microarrays," in their search for them. DNA chips are sets of probes that can scan an entire genome to compare which genes are being used to produce RNA molecules in different types of cells. This makes it possible to compare cancerous and healthy cells, which may reveal links in the Notch information chain that are only used in tumors. This might also reveal new ways to shut off their blood supply.

Understanding the development of the circulatory system will be necessary to make advances against a much wider range of health problems, particularly cardiovascular diseases, the world's greatest killers. Many of these diseases arise because blood vessels become blocked, which prevents supplies of oxygen and nutrients from reaching the heart, the brain, and other organs. Sometimes this happens because fats or proteins accumulate clumps that cannot be dissolved. Another disease, vasculitis, occurs because of damage to the cells that line the vessels or other types of problems. One form of vasculitis arises because the body builds *antibodies* against white blood cells. Usually these cells slip through the lining of blood cells to look for invaders such as viruses or bacteria; when they find one, they explode like miniature hand grenades, releasing substances that break up other cells. In ANKA-vasculitis they explode too early and damage the lining of blood vessels. This results in blockages that starve organs of blood and often lead to death.

Heart attacks are also fatal because of oxygen starvation, in this case to cardiac muscle, and they are frequently linked to blood vessel damage. This usually has something to do with the characteristics of coronary arteries. While the heart supplies blood for a vast circulatory network, it has to be supplied with blood itself. Pushing blood into such a massive muscle is a hard job, handled by special coronary arteries. Too much pressure would be necessary to do this when the heart contracts, so blood only penetrates deeply into the muscle when it relaxes between contractions.

The supply comes from arteries in the epicardium, the outer heart layer. When these vessels are healthy, they self-regulate the flow and pressure needed to supply the organ. They are a weak point in the system, however. One reason is that they are narrow, which means that they become clogged in conditions such as artherosclerosis, and another is that they are the only source of blood that the heart has. If the flow is interrupted even briefly, the muscle is starved of oxygen, and damage to the tissue happens very quickly. There is a feedback system that makes the heart begin beating extremely quickly, but if oxygen deprivation goes on very long, the result will be a massive heart attack. If this is not fatal, the best possible outcome is usually severe damage.

THE DIGESTIVE SYSTEM

Blood provides the body's cells with the nutrients they need after it obtains them from food and air, which are absorbed in the gut and lungs. The respiratory and digestive systems arise from a single tube that begins to form very early in human development, about 16 days after fertilization of the egg. Different regions of the tube or buds that sprout along its sides eventually become the stomach, lungs, liver, and several other organs.

The tube develops when a long band of endoderm cells pushes its way into the belly region of the embryo. It begins as a sheet, but soon it folds inward near the head and at another point near the tail, then becomes pinched off. The resulting tube gives rise to most of the organs responsible for digestion and respiration. It has multiple layers, like a pipe wrapped in several sheets of insulation. The inner layers form from cells of the endoderm, which are then wrapped in muscle from the mesoderm. At different rates of growth cells along the tube pull the head region down toward the chest, giving the embryo a curled, shrimplike shape.

At first, the tube has three major regions—the foregut, midgut, and hindgut—which quickly specialize further. The top end alternately grows, creating bulges, and is squeezed to create a row of sacs arranged in pairs. These are called the pharyngeal pouches (named for the pharynx, the tube in the throat that later connects all of the organs that arise here). The first pair develops into the two middle ears. The next becomes shaped into the tonsils, and the thymus forms from the third pair of pouches. Between the fourth and lowest pair of pouches arises the respiratory tube.

The thymus is a main actor in the immune system because it hosts cells that develop into *T cells,* a specialized type of white blood cell. They have several roles to play in defending the body against invaders: They attack other cells that have been infected with viruses; they release small molecules called "cytokines," which attract other immune cells; and they maintain a "memory" of earlier infections so that the immune system can respond immediately if a virus comes back for a second try.

Regions of the tube below the pharynx develop into the esophagus, the stomach, and the small and large intestines. They receive most of their developmental signals through contact with the mesoderm, which activates genes that will be specific to each organ. Some of the programming occurs even before the endoderm tube has formed. Working with chick embryos, Susumu Matsushita of the Tokyo Women's Medical University and Sadao Yasugi of Tokyo Metropolitain University showed that genes specific to particular organs are already active in endodermal tissue even before it folds into the abdomen of the embryo. For example, the presence of Sox2 protein marks the esophagus and stomach, while CdxA is produced only in the intestine. These molecules can already be detected in the endoderm.

The digestive tube and its organs are not the result of a one-way transmission of signals from the mesoderm to endoderm. As with arteries and veins, cells that have begun to differentiate can transform their neighbors into helpers that they will need later in development. While the endoderm specializes, it sends back signals that create different types of mesoderm tissue. The Sonic hedgehog molecule is thought to be one of these signals. Usually it is produced only in certain places along the gut, and what it does depends on other molecules that are active in each location. In an experiment carried out by Drucilla Roberts, Devyn Smith, and Clifford Tabin, developmental biologists at Harvard Medical School, a virus was used to deliver Shh to the entire digestive tube of chick embryos. This showed that Shh does not have the same effect on all of its neighbors, which, in turn, means that cells have already received developmental instructions before the tube forms. When the same experiment was tried with a signaling molecule called Hoxd-13, it made the developing midgut begin to look more like the hindgut.

LIVER, PANCREAS, AND GALLBLADDER

The pancreas, liver, and gallbladder begin as a second single tube that grows off the small intestine, just below the stomach. A signal from the mesoderm tells some of the endoderm cells to create a stem that dead-ends in a bulb. That bulb becomes the liver, and the stem leading to it sprouts smaller buds that become the pancreas and gallbladder.

Many genes needed to build the liver are active in several places along the digestive tube, but it only develops in one place thanks to signals from other tissues. In the 1970s Nicole Le Douarin of the University of Nantes in France showed that one of these signals, probably the molecule FGF, came from the

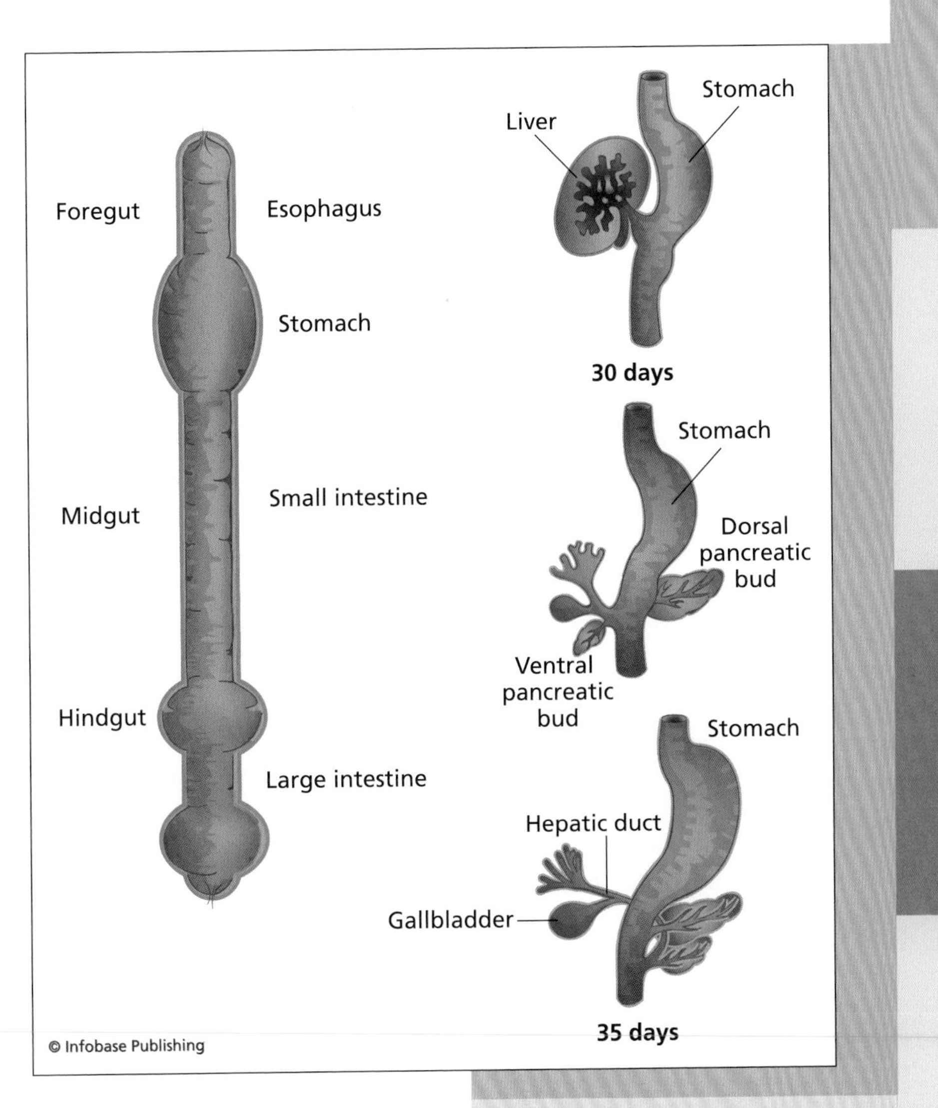

The esophagus, stomach, liver, pancreas, gallbladder, and intestines all form from a single tube through a process of growing, bending, and the formation of buds and stems.

mesoderm cells forming the heart. Then in 2001 Kunio Matsumoto, Hideyuki Yoshitomi, and Kenneth S. Zaret of the Fox Chase Cancer Center in Philadelphia discovered that another signal was necessary. They were studying how the removal of a molecule called Flk-1 affected mice. One result was that the mice failed to develop the endothelial cells that line blood vessels. Another was that tissue that had begun to specialize into liver cells failed to take the next steps. The discovery was possible owing to a new method that the scientists had invented to extract liver buds and follow their development in the test tube.

Matsumoto and his colleagues noticed that in healthy mice these endothelial cells surround the liver bud. They define the area that the bud will grow into; without them the organ does not form, and there is no migration of blood cells into the region. This is similar to the way mesoderm and endothelial cells interact in forming the heart, and the scientists believe that endothelial cells may have a similar impact on the formation of additional organs and tumors. It may be possible to prove this using the new techniques from the Matsumoto lab.

As the liver forms on the abdominal side of the digestive tube, two other buds are taking the first steps toward becoming the pancreas. The function of this gland is to secrete digestive juices into the small intestine. It also releases hormones such as insulin into the bloodstream. Some forms of diabetes arise because the pancreas cannot do this, and in many cases the reasons lie in its development. Pancreal development is like that of the liver, with two exceptions. It is built from two buds (one on the ventral side, near the stomach, and the other toward the back). The bud in the back comes in contact with cells of the notochord, and the other with the heart, on the ventral side.

The result is to shut down the Shh molecule only in this region of the gut. Sonic hedgehog was introduced earlier in the book as one of the major molecules used to communicate between the endoderm and mesoderm, leading to differences in the front and back regions of the brain—and also for its role in the beginnings of the digestive system. To build the pancreas its activity has to be blocked. The reason may be that it permits other signals to arrive from developing blood vessels. (Blocking contact between vessels and the bud also prevents the pancreas from forming.) But if things develop normally, their contact causes cells of the future pancreas to begin to produce the gene-activating molecule Pdx1. Liver cells do not make this protein, but in 2001 Marko Horb, a developmental biologist at the University of Bath in Great Britain, carried out an experiment that forced the cells to do so in the embryos of frogs. The result was that the liver was transformed into a second pancreas.

The gallbladder forms between the bud of the pancreas and the liver; its job is to store and release bile produced by the liver. The organ secretes this yellowish liquid into the small intestine, where it is needed to process the fats in food. Fats cannot be dissolved in water, so bile is needed to break them down and release hormones and other nutrients that they contain. The gallbladder develops under the influence of the same molecules that create the liver, but some small differences keep it from becoming a second liver. In 2002 Frédéric Clotman and Frédéric Lemaigre's lab at the Catholic University of Louvain, Belgium, found a molecule that helps distinguish the two tissues: a transcription factor called HNF6. Removing it from a strain of mice made the gallbladder (but not the liver) disappear; it also rerouted the liver's drainage system.

The rest of the digestive tract forms from regions of the gut below these organs. The midgut gives rise to most of the intestines. In adults this tube is usually about 20 feet (6 m) long and plays the main role in digestion. Food has already been processed before it arrives. First the stomach has broken down proteins into their building blocks, amino acids, for recycling. The food is then pushed out of the stomach by a muscle. It enters the intestines, where it is forced along by pulsing contractions from the muscles that line the tube. First, it enters the small intestine, where it is broken down further by protein-chopping molecules that have entered from the pancreas.

The lining of the small intestine is covered with tiny, fingerlike protrusions called "microvilli." Its wrinkled structure allows it a much larger surface to come into contact with food and draw more nourishment. The first 10-inch (25 cm) section absorbs iron. The second segment is about 8.2 feet (2.5 m) long and takes in nearly all of the other nutrients the body needs, except for vitamin B_{12}, which is absorbed in a region at the very end of the small intestine. Water and fats are extracted from food all the way along.

The hindgut gives rise to the colon (large intestine), rectum, and anal canal. By the time food reaches the colon, the nutrients have been largely extracted. The colon extracts and stores most of the remaining water; it also absorbs nutrients produced by resident bacteria, then passes the leftover waste from the body.

The colon has long been known to host hundreds of species of bacteria, but little is known about most of them because they cannot be grown in the laboratory. Some of the gaps are being filled in by new techniques called "mass genomics," which capture and analyze free-floating genes in an environment rather than taking them from organisms. In 2002 Mya Breitbart and Forest Rohwer of San Diego State University used these methods to discover traces of more than a thousand unknown species of viruses and bacteria in human wastes. Many of these have probably coexisted with the human species for long periods of time and very likely have important—but so-far unknown—functions in the body.

LUNGS

On the one hand, it may seem strange that the organs needed to breathe and eat begin in the same place; everyone knows the awful sensation of getting food down the "wrong pipe." On the other hand, in land animals both systems start at the mouth and throat. Very early in development they are one tube, but soon the breathing apparatus starts as a new branch off the digestive tube, later becoming a second passageway. In 2004 Jun-Ichi Sakiyama and Atsushi Kuroiwa's laboratory at Nagoya University in Japan discovered that a molecule that helps build hind limbs was also crucial to the first steps in the development of the lungs. The molecule that is responsible is a type of transcription factor called Tbx. (This molecule appeared earlier, as a protein with a dual function in heart and limb development.) By forcing the wrong cells in the gut tube to produce Tbx, Sakiyama and her colleagues produced animals with extra lung buds. They could also do the opposite; by blocking Tbx, they stopped the formation of the trachea and lungs.

The initial bud grows and stretches into a respiratory tube that is long and straight at the top, where it splits off the digestive tract. Below it branches again and again into ever smaller tubes that finally dead-end in tiny, balloon-like sacs called "alveoli." They have extremely thin walls that allow oxygen to pass into the bloodstream and carbon dioxide to move out. In the 1970s Norman Wessells of Stanford University wanted to find out what made the two ends of the tube different—in other words, what kept the top part from branching like the bottom. He removed early lung tissue from mice embryos and put different parts in contact with different types of cells. Those that encountered mesoderm cells from the region of the trachea remained straight, while the rest branched off in the normal way. This suggests that left on their own, the cells of the trachea go on to develop branching lung tissue; the process has to be stopped at the top of the tube through a signal from the mesoderm cells. In 2001 Wei Shi and David Warburton at the Children's Hospital Los Angeles Research Institute discovered that a familiar molecule was responsible for branching: Bmp4. (This protein was introduced in chapter 2 because it stimulates formation of the head.) Branching does not happen in the upper respiratory tract because Bmp4 is blocked by a molecule called "Gremlin."

Lungs only become fully functional late in development. They do not have to work until birth, but then they have to do so immediately. This can cause serious problems for premature babies because their alveoli are coated with a sticky substance that glues them together, preventing oxygen and carbon dioxide from passing through. Babies born after the 34th week of life seldom have this problem because by then the cells of their alveoli have secreted fats that dissolve the "glue." This frees the surfaces of the sacs and the atoms can now pass through.

One reason that smoking causes such serious problems for the lungs is that over the long term, it gums up the alveoli. Cigarette smoke and other toxic fumes contain large molecules that plug up the small holes in these membranes, the way that dirt or fat plugs up a window screen. The effects of smoking on other parts of the body are discussed at the end of the book.

BONES

Many of the body's organs arise from sheets of cells that bend into tubes or expand and push against other tissues. Yet, a bit surprisingly, some of the strongest and most structured parts of the adult body—bones and muscles—arise from disconnected groups of cells that migrate into dense clumps that condense and then fuse.

There are two major types of bone formation. One is responsible for producing nearly the entire skull, beginning when cells from the neural crest condense and secrete proteins that latch onto calcium atoms. This forms a scaffold that some cells become locked into; they develop into needle-like shapes. Other cells are attracted to the surfaces and deposit more layers on top, enlarging them.

The rest of the body's bones and muscles originate from loose tissue in the mesoderm. Throughout development groups of cells alongside the neural tube are released to form separate structures called *somites* (see image on page 54); later, their cells will be released to migrate through the body and take on several functions. Somites are ball-shaped clusters of cells that become wrapped up in a skinlike layer of epithelium. At first, the regions of a somite and the fate of its individual cells are not determined. They can develop into many things: cartilage (and later bone), the dermis (the main layer of the skin), muscles of the body wall and limbs, or back muscles. At an early stage the somite can be rotated, and cells will develop properly into all the proper types. But as different parts of the ball receive signals from neighboring tissues, each somite develops four specialized regions, one region devoted to one of these four types of tissues.

Cells on the ventral (abdominal) side receive signals from the notochord that make them let go of each other and migrate toward the neural tube. The signal involves a molecule called Pax1, and it starts cells along a developmental path that creates the bones of the arms, legs, vertebrae, and ribs. Released from the somite, cells migrate out to the places where limbs and the other structures form. Once on site, the cells collect and condense. Edwina Wright and Peter Koopman of the University of Queensland in Australia discovered that they then begin to produce a gene-activating molecule called Sox9, which is vital to the development of the skeleton. A mutant version of Sox9 leads to a rare genetic disease in which most of the body's bones are deformed during development. Embryos with the mutation usually die at birth or shortly afterward.

If the mesoderm tissue is programmed correctly by Sox9 and other proteins, it builds a "first draft" of the future bone out of cartilage. This dense substance is made of cells, protein fibers, and a gel-like material called "matrix." In the center a group of cells called "chondrocytes" inflates to a huge size. They begin to produce molecules that stick to calcium, and this makes the cartilage mineralize. They also stimulate the formation of blood vessels. At this point they die, leaving gaps that act as passageways for vessels that will later supply blood to the inside of bones. First, this inner space has to be hollowed out by cells that invade from the outside.

The compartment has a crucial function in adulthood. It becomes filled with marrow, the main source of stem cells used to make new blood cells. These stem cells exist before either the bone or its inner compartment; in the very early embryo they form around major arteries, then migrate to the liver, and at birth move to the bone marrow.

Bone diseases reveal the great flexibility of mesodermal cells during development. Often a single mutation transforms one type of tissue into another. In people who suffer from fibrodysplasia ossificans progressiva (FOP), too much tissue—including muscles, tendons, and ligaments—becomes bone. In 2006 Eileen Shore and Frederick Kaplan's lab at the University of Pennsylvania School of Medicine linked FOP to a molecule called ACR1, which receives signals from Bmp proteins. In bone Bmp proteins stimulate the production of molecules that mineralize cells. Interfering with the molecules themselves or the pathway by which their signals are transmitted to genes can lead to animals without bones at all. The signal should only appear in the right tissues at the right times. In FOP, however, a mutation in ACR1 freezes the pathway in the "on" position, so too many types of cells receive the signal to transform themselves into bone.

SKELETAL MUSCLE

Muscle tissue comes in three types. Smooth muscle lines the blood vessels and the walls of organs such as the stomach and intestines and is responsible for involuntary movement. Cardiac muscle forms the heart. *Skeletal muscle* is the type familiar to most people and is the type that will be discussed here. It is connected to the bones and places the motion of the body under the conscious control of the brain. It is composed of bundles of stripelike fibers that can be seen under the microscope, which earned this type the name striated muscle. In skeletal muscle the fibers form long, straight rows; cardiac muscle is similar, but the fibers branch off at angles.

A fully developed human has more than 600 distinct skeletal muscles, made of unusual cells that have fused together in long fibers. The fibers are woven together in bundles that usually stretch from a more central part of the body to a more distant point, normally a joint or a movable tissue such as the eye. Assembling the fibers and hooking them up to their proper locations in the body are complex processes that begin in the early embryo. Scientists are keenly interested in identifying the molecules and processes that underlie muscle formation because they hope to find explanations for developmental defects in the tissues and the causes of diseases in which they break down.

Muscle development is known as *myogenesis,* and like the formation of bone and skin, the process begins in somites. Some of the cells in these structures are triggered by a developmental signal called Wnt to begin producing MyoD and Myf5 proteins, which launch their specialization into muscle. As this happens, the somite flattens to become a sheetlike structure called the "dermamyotome," with developing cells clustered underneath. The bottommost cells will form cartilage and bone; sandwiched in between are those destined to become muscle.

One very active area of research is to decipher the signals that tell cells from particular somites to migrate to specific places in the body. In 2006 Carmen Birchmeier's laboratory at the Max Delbrück Center for Molecular Medicine in Berlin, Germany, showed that a signal called Lbx1 was necessary to direct developing muscle cells to the limbs of mice. When they knocked out the gene, the limbs did not properly develop because they had no muscle. Cells did, on the other hand, migrate to other parts of the body. Each region and tissue likely relies on its own set of signals—or its own combination—because otherwise the wanderers would get confused. But many of the cues have not yet been identified.

The structure of skeletal muscle itself was a puzzle to scientists until about 1960. The tissue is composed of mysterious cells that contain long fibers and several nuclei. Cardiac cells are similar; their nuclei divide to create these structures. Detailed studies with the electron microscope, carried out by Irwin Konigsberg of the University of Virginia, showed that separate cells were fusing to make skeletal muscle. Myoblasts, the stage between stem cells and fully developed muscle, stop dividing and secrete a protein called "fibronectin," which acts as a sort of intracellular glue. Without it fibers never develop. The next step is that neighboring cells line up into rows. This is managed by surface molecules called "cadherins" that recognize receptors on other muscle cells and dock onto them.

In 1995 Takako Yamagi-Hiromasa and other members of Atsuo Fujisawa-Sehara's lab at the National Institute of Neuroscience in Tokyo found a molecule that was essential for this process. The team began with the hypothesis that many different types of fusion—between egg and sperm, viruses and cells, and the building blocks of muscle—might be based on a similar process and some of the same molecules. They had some evidence: They had found that part of a protein from the rubella virus strongly resembled part of a mouse protein called fertilin-alpha, which was necessary for the fusion of egg and sperm. They began searching for other proteins that might contain the pattern and found yet another molecule called meltrin-alpha, produced in the muscle cells of embryos and newborn mice. Blocking this protein stopped the fusion of myoblasts into fibers.

Muscles' ability to expand and contract comes from microscopic "pistons" called "sarcomeres." They are stacked up end to end in long rows. The piston itself is a dense rod made of the protein myosin. It is housed in a shell of thinner fibers made of

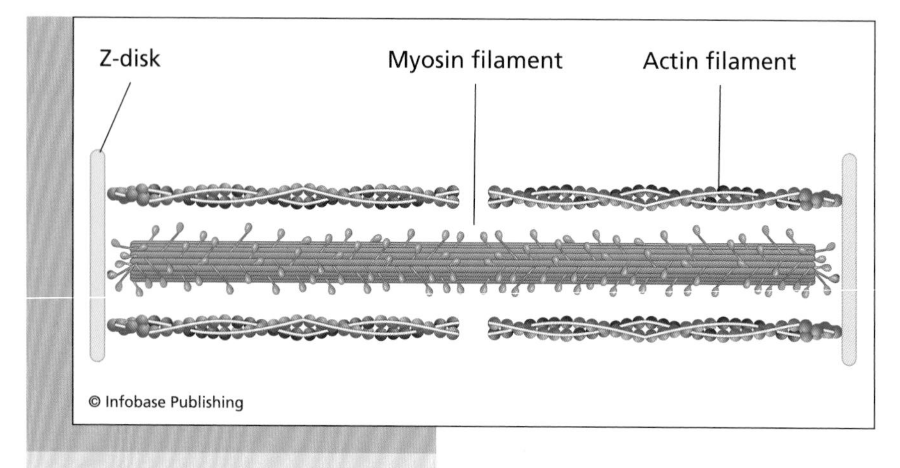

Skeletal muscle can contract and relax due to pistonlike structures inside muscle cells (sarcomeres). They are built of protein fibers that slide by each other in a ratchetlike movement. A springlike protein (titin, not shown) anchors the fibers to a thick band of proteins at either end (the Z-disk).

actin. This protein has small "feet" on its surface that allow it to crawl along the myosin, sliding the piston in and out. When this happens in a coordinated way along the whole fiber, the muscle contracts and relaxes. Building the sarcomere seems to be partly the work of a huge molecule, titin, the largest protein produced by human cells. It is anchored in the end of each sarcomere—a caplike region called the "Z-disc"—and spans half the length of the piston. Another copy of titin stretches toward it from its anchor in the opposite end. Along the way the molecules come in contact with all of the components of the sarcomere. Titin acts like a huge spring because some of its regions unfold when placed under stress, then snap back together again when the tension is relaxed. Some regions of the molecule appear to act as sensors to detect the amount of tension placed on the structure.

Mutations in titin or other sarcomere components have been found in patients with inherited forms of some types of heart disease. In 2007 Michael Gotthardt's lab at the Max Delbrück Center in Berlin showed that a "signal receiving" module in titin helps muscles contract. If it is defective, the structure of sarcomeres begins to dissolve. Gotthardt and his colleagues showed that this happens because the cells cannot properly handle calcium; this element is essential in coordinating the activity of muscle cells. Other mutations in titin block its springlike activity, leading to sarcomeres that are too small and not as flexible as they should be.

To a limited extent adult bodies can also build new muscle. This happens when tissue suffers damage or when a person exercises. The process is quite different from what happens in the embryo because adults no longer have somites. Instead, fresh cells are obtained from a more local source: a type of stem cell called a *satellite cell,* found at intervals on the surface of skeletal muscle fibers. Signals tell the cells to complete their development and move into the fibers. Until that time, like other adult stem cells, they need to be actively maintained by molecular signals until they are needed.

Problems with this process of renewal seem to lie at the heart of some degenerative muscle diseases. In 2007 Birchmeier's lab in Berlin showed that satellite cells need a protein called RBP-J to interpret the signals that preserve them. RBP-J acts as a sort of "gene brake" that blocks the activity of developmental genes. It should only be released when cells receive a signal from the protein Notch. Without RBP-J, Birchmeier discovered, the satellite cells lose their stem cell status and start to differentiate. When that happens, the body exhausts its supply and can no longer build muscle. Any tissue that is lost will not be replaced, and the person will suffer from progressive muscle degeneration. This happens as a normal part of aging. The lab hopes that their work on the signals will help explain why and possibly give some insights into how to prevent the process.

SKIN

Skin is made of several layers of specialized cells that come from different parts of the embryo. The outermost coat, the epidermis, consists of five major sublayers, and each of these is composed of thinner sheets of cells. They make the body watertight and protect it from minor injuries. They come from the ectoderm, the outermost layer of the embryo that forms during gastrulation (see chapter 2). The epidermis begins as a one-cell layer that becomes thicker as its cells divide. Its outermost region is composed of 25 to 30 layers of dead cells that flake off; new cells are created below and move upward to replace them. At the bottom of the epidermis reside melanocyte cells, which produce pigment proteins. The pigments are passed along in small sacs to a middle layer of *keratinocytes,* the cells that determine a person's skin color.

Cells in the lowest level of the epidermis reproduce to replenish the upper layers. Normally it takes about eight weeks for a newly made cell to move up to the surface, where it stays for two more weeks before being shed. In people who suffer from psoriasis, these cells flake off much faster, often after only two days at the surface. In 1989 James Elder and members of John Voorhees's lab at the University of Michigan at Ann Arbor showed that a protein called TGF-alpha was responsible for these problems. They found that in people with psoriasis, TGF-alpha is produced at unusually high levels in epidermal cells, and it makes them divide too rapidly. In 1991 Robert Vassar and Elaine Fuchs of the University of Chicago produced a strain of mouse in which TGF-alpha was permanently switched on in the epidermis. The animals developed a scaly skin with far too many layers of dead and shedding cells.

Below the epidermis is a much thicker layer of skin, the dermis. It comes from the ball-shaped somites near the neural tube, which also generate bones and muscle. Signals from the neural tube launch a genetic program that causes some of the somite cells to migrate to the inner side of the epidermis and create the dermis. Gilat Brill and Chaya Kalcheim's lab at the Hebrew University–Hadassah Medical School in Jerusalem identified neurotrophin-3 as one protein in the signaling system. In 1995 they used antibodies to block the molecule in chick embryos and found that somite cells failed to migrate and form dermal tissue.

The dermis holds blood vessels, sweat glands, and nerve endings—all vital to the skin. Fibers made of proteins anchor the cells

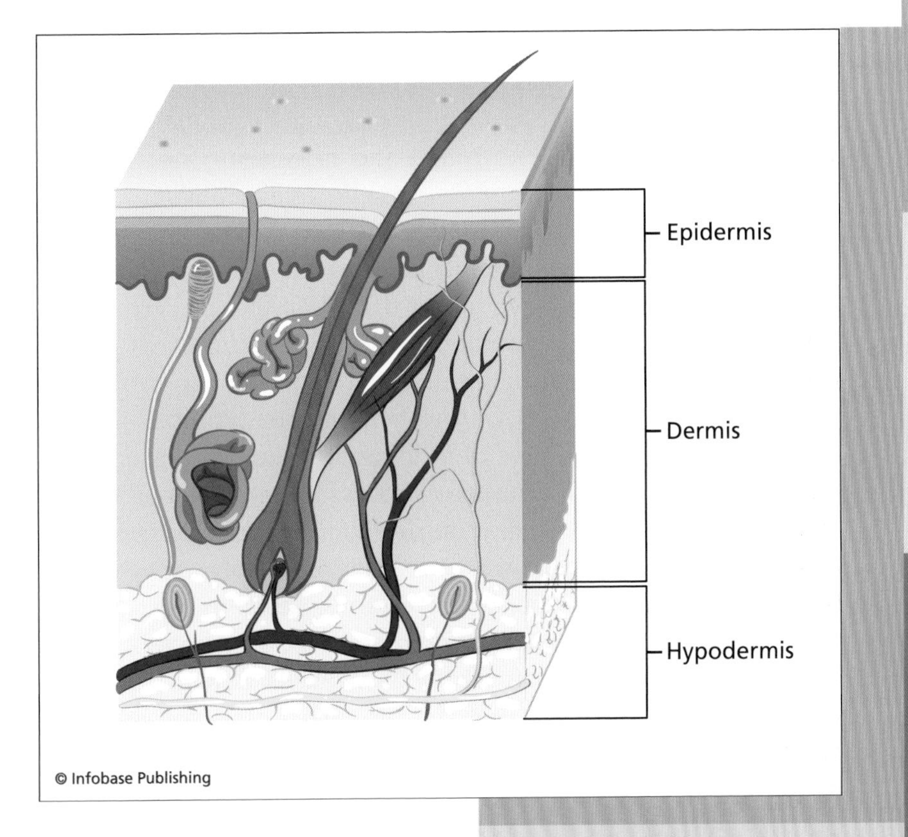

Skin is made of many layers of cells that originate in different parts of the early embryo. The dermis, for example, arises from ball-shaped somites.

to one another horizontally and vertically, which allows the skin to stretch without becoming "unglued." The dermis rests on a "mattress" of other cells, the hypodermis, which fastens it to muscles and inner tissues of the body.

The growth of these layers has to be coordinated and carefully controlled during development so that there is enough skin to cover the growing body. Throughout a person's life new skin also has to be produced to heal small wounds. Humans and mammals have such high blood pressure that they need a fast-acting emergency system to seal off the skin when there is a cut, but there is a drawback to the system: It leaves a scar, a dense mass of fiber and other material that prevents new cells from moving in and growing new skin. Embryos and small children have a much better capacity to regenerate skin without scars, but this ability is lost over time, which is why wounds are more likely to leave scars in adults. Identifying the molecules that switch regeneration on and off—and how they function differently in adults and embryos—might give scientists an idea about new ways to treat wounds.

Studies of skin development are also providing insights into skin cancers, which arise in either the epidermal or dermal layers of skin. Often they are the result of a mutation that causes cells to divide at the wrong time. These cancers bear similarities to both development and wound healing. For example, tumors frequently form at the sites of injuries or wounds that require a long time to heal, as if a healthy process has started but cannot be shut down again. And new studies with DNA chips and other techniques show that genes that are normally only active in embryos or in inflammations are often reawakened in cancer cells.

Most of the tumors can be seen at an early stage as unusual spots, moles that begin to grow or wounds that do not heal in an expected way. If caught at an early stage, the majority can be removed by surgery. If untreated, some tumors become malignant melanomas that are deadly because they spread to other tissues and begin to grow there, disturbing the function of important organs.

In recent years the frequency of skin cancer has risen dramatically in the United States, partly due to the popularity of sunbathing and increasing amounts of ultraviolet (UV) light that penetrate the atmosphere. This type of radiation is known to cause mutations, and heavy exposure to sunlight (particularly in childhood) has been linked to a high risk of developing the disease later in life. Fortunately, sunscreens, protective clothing, and staying indoors during hours of most intense sunlight can greatly reduce exposure to these dangerous rays.

LIMBS

In the 18th century a French doctor named Pierre-Louis Maupertuis (1698–1759) became fascinated by a family with an unusual

The major-league baseball pitcher Antonio Alfonseca has inherited a form of a gene that gives him six fingers on each hand and six toes on each foot.

characteristic: extra fingers on their hands. Maupertuis was intrigued by the fact that not all members of the family had them and began a study of the pattern by which they were inherited over four generations. It was the first scientific study of human heredity, and a century and a half after his death the same methods would be used to prove that human genetics follow the same rules as heredity in animals and plants. Having an extra finger or toe might make it hard to find a pair of gloves or shoes, but otherwise it probably would not do any harm. In fact, it might have some advantages: Antonio Alfonseca, a baseball pitcher for the Florida Marlins, has six fingers on each hand, and it did not give him problems when he played in the 2003 World Series. So, why is the mutation that gives people extra digits so rare?

One part of the answer might lie in the precision needed to create arms; after all, the body has to build two that are almost exactly alike. That is a more general problem during development:

Humans have overall bilateral symmetry (mirror-image left and right sides), which evolved in an animal at least 600 million years ago. *Urbilateria,* as this organism is called, came equipped with genetic programs that tightly controlled and coordinated the growth of the two sides so that they became the same size. Those programs were so important that they were maintained with high accuracy in all of the animal's descendants.

Studies of limb development are a very good example of how developmental biologists search for the genes responsible for particular traits. The genes and tissues can usually be manipulated without causing serious developmental defects in an animal, and the effects of a particular molecule can easily be detected just by looking at the limb.

Limbs begin as cells in the mesoderm. Their muscle comes from tissue called the "lateral plate" mesoderm, which begins in the back alongside the notochord and stretches forward toward the abdomen in a sheet, eventually joining up. The tissue passes the somites near the notochord, and it signals them to release some cells. Those migrate to the sides and create the limb field, a bump where the shoulders and hips will be. In the early 20th century Samuel Detwiler and Ross Harrison at Yale University in New Haven, Connecticut, showed that removing this tissue from a salamander and transplanting it somewhere else could cause a limb to grow in another place; in other words, the cells of the limb field contain all the information and produce all the signals needed to form a complete arm or leg. Removing all the cells of the field completely blocked the development of a limb; if some were left behind at an early stage, they could rebuild the field and the limb. Transplantations could even work between different species—to an extent. Placing the limb bud of a mammal into a chick would make the body build a leg in that place (but that of a chick, not a mouse!). If the limb bud is cut in its early stages, two limbs may grow in place of one.

Mammals can regrow limb buds that have been damaged or lost up to a certain stage in embryonic development. They later lose this regenerative capacity, but other organisms keep it their entire lives. If some species of salamanders or other amphibians lose a leg, a new one grows in its place—once again starting as a limb bud and going through "embryonic" stages of development.

Over the last two decades many of the key molecules that guide the formation of the bud and its further growth have been identified. The powerful developmental Hox molecules play an important role. A protein called Hoxc6 determines the position of arms (forelimbs); their limb buds develop at the headmost location along the neural tube where Hoxc6 is produced. Hox genes are often activated by vitamin A (retinoic acid), which is also needed in specific locations to launch the development of a limb. In 1996 Thomas Stratford of King's College in London showed that drugs that stopped cells from producing retinoic acid also blocked the development of limb buds. The substance appears to play an equally important role in regeneration. Priyambada Mohanty-Hejmadi, a developmental biologist at Uktal University in Bhubaneswar, India, showed this in experiments with tadpoles. He removed their tails and applied retinoic acid to the wound; several legs grew from the site.

A limb bud begins to form when cells in the lateral plate mesoderm secrete a protein called FGF10. At first, this molecule is produced in a large area of the mesoderm, but it fades away again in most places. It needs to be kept active only where limb buds should be built, which is the job of another powerful developmental protein, Wnt. Next, transcription factors determine whether the bud becomes a forelimb or hind leg. The proteins Tbx4 and Pitx1 were known to be linked to development of the legs, and a related molecule called Tbx5 is made in the area where arms, wings, or forelimbs form. In 2006 April DeLaurier of the National Institute for Medical Research in London developed a strain of mouse whose forelimbs produced Pitx1; as a result, they took on the form of hind legs. What triggers the appearance of different molecules in the two places is not yet known.

Cells in the bud continue to produce FGF10, which causes the cells at the forward edge of the bud to create a structure called the "apical ectodermal (AE) ridge." This structure remains at the front as the limb grows away from the trunk of the body, like the crest of a wave. The ridge secretes another molecule, FGF8,

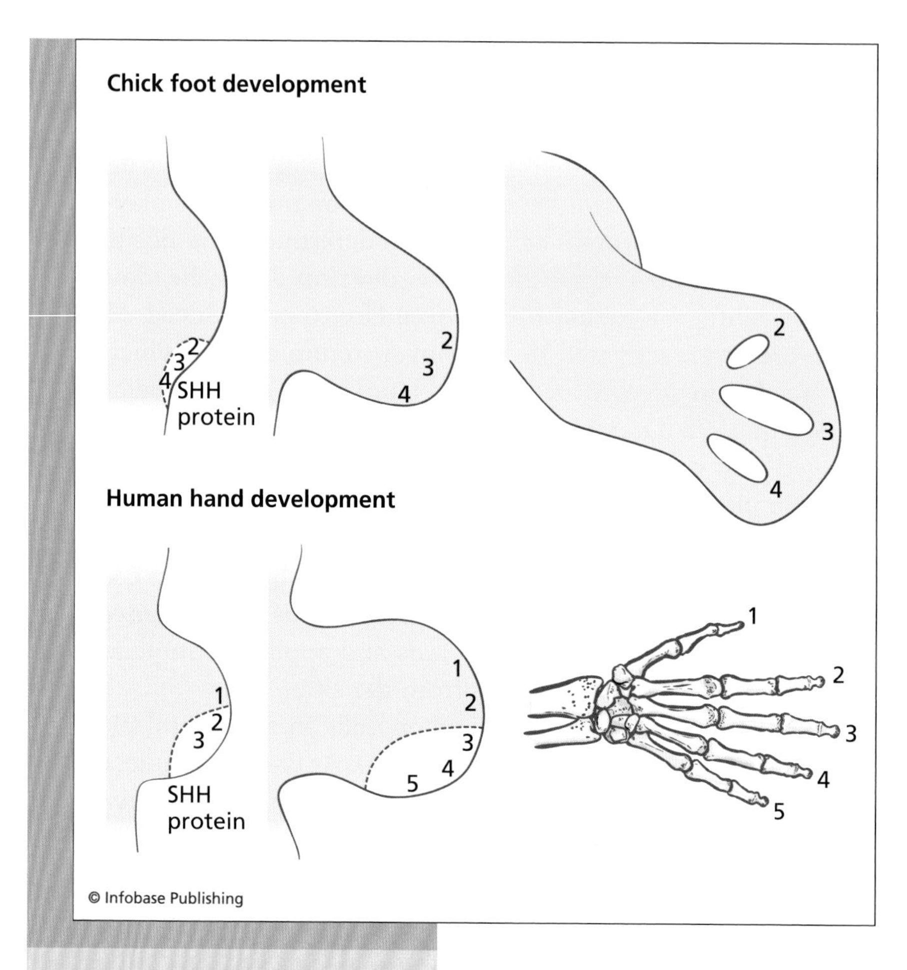

Limbs begin as bumps that grow and take on structure due to molecular signals. The length depends on communication between the forward edge and the tissue behind it. The number and position of digits (1–5 in the human hand, 2–4 for the second to fourth chick toes) depend on the source of a signal called Shh and the amount of time it is transmitted. The numbers track specific regions that will develop into specific digits.

which makes the cells immediately behind it divide at a fast rate. Cells in this region, called the "progress zone," respond to FGF8 by creating more FGF10, which makes a feedback loop: The cells that grow maintain the ridge, and it keeps giving them instructions to divide.

Even before scientists had identified any of these molecules, transplantation experiments carried out by John Saunders in 1948 at John Hopkins University in Baltimore, Maryland, began to reveal some of the principles behind limb development. Saunders wanted to know what controlled growth along the longest axis of the limb, the shoulder-to-fingertip direction. He discovered that removing the apical ectodermal ridge at different times had different effects. If it was removed very early, only the uppermost bone of the arm or wing formed. Taking it away a little later produced arms with the upper and forearm bones but no hand. This raised a puzzle: Which cells stored the instructions?

Further transplantations have provided some of the answers. Scientists have moved tissue—the ridge, the zone, or both—into other limb buds at different stages. This either stops growth, blocks the development of particular parts, or adds extra structures in the wrong places. These experiments have shown that cells receive instructions for their places in the limb very early—possibly even as the bud forms. Once the programming has begun, the cells move to the right place, talk to their neighbors, and grow to become the right length for each part.

Just as Hox genes help set up the head-to-tail axis of the body, they work to create the right structures in the right places in limbs. Once again, a cluster of Hox genes becomes activated in sequence; the first gene on the chromosome is used to build the tips of fingers, toes, or wings, and the structures that follow are determined by genes in the proper order. These ensure that the structure makes architectural "sense" (for example, by not growing fingers in the middle of arms).

In the late 1950s a German pharmaceutical company, Grünenthal, marketed a sedative called thalidomide (contragen) that turned out to disrupt this process. If pregnant women took the drug between the 20th and 36th day after conception, their children were likely to be born with serious developmental defects of the limbs. Many women had used thalidomide to ease the symptoms of morning sickness, but an alarming rise in the number of children with birth defects raised suspicions, and the drug was taken off the market in 1961. The effects of the drug on development were not truly understood until 2007, however, when a study by Jürgen Knobloch and Ulrich Rüther at the Heinrich-Heine University in Düsseldorf, Germany, revealed that the drug causes an overactivation of Bmp molecules and the genes that they activate. This leads to the death of certain types of cells in crucial parts of the limb bud.

Limbs are asymmetrical in all three dimensions, easiest to see by looking at a hand: The fingertips are different from the wrist, the thumb is different from the little finger, and the top is different from the palm. (Entire arms, wings, and legs are likewise asymmetrical in the three directions.) Explaining limb development means finding the molecules responsible for making these differences. Interactions between the AE ridge and the underlying cells explained part of the shoulder-to-fingertip development, but what about the other directions?

The difference between the palm side and the knuckle side arises because of signals exchanged between the ectoderm (skin) and the mesoderm tissue that will build the bones and other structures of the hand, foot, or wing. Skin on the back side secretes a molecule called Wnt7a, which goes on to activate a molecule called Lmx1 in the mesoderm cells. Without Wnt7a animal paws develop footpads on both sides. If Lmx1 is produced—inappropriately—on the palm side, both the top and bottom look like the top. Without Lmx1 at all, limbs mix up the two sides. In rare cases this protein is mutated in humans; their hands have no nails, and their legs do not develop kneecaps.

As the limb bud grows, signals come from different sources, at different intensities, in new combinations. This establishes a three-dimensional coordinate system that each cell can read to discover its exact location. Small groups of cells—sometimes even single cells—receive a unique recipe of signals and respond in unique ways. They produce new molecules that change their fates and those of their neighbors.

One of the most interesting topics in limb development is the asymmetry that develops between the thumb and little finger sides. Here, too, the AE ridge seemed to play an important role. In the late 1960s Saunders and Ph.D. student John Fallon at Marquette University in Wisconsin discovered a cluster of cells that appeared to control the direction of thumb to little finger. These cells grew in a small "zone of polarizing activity" (ZPA), located at the junction where the limb bud connects to the trunk of the body. Removing cells from this region of an animal and transplanting them to another limb bud, to the opposite side, created a double set of digits there. A special feature of the ZPA seemed to be that its cells produced Shh, a Hox protein.

Robert Riddle and Cliff Tabin of Harvard Medical School proved that it contributes to the formation of limbs (as well as the brain, the heart, and the gut) in 1993. Riddle used a virus to insert Shh into other types of cells. When these cells were transplanted into another limb, the result was once again extra digits. Further work has shown that normally the AE ridge secretes an FGF protein that activates Shh. As with most developmental questions, solving one immediately raised another: Only some of the cells that received the signal responded, probably due to instructions passed along by other signals.

The timing of Shh activation is very important in determining how many digits form. For many years Nadia Rosenthal (who wrote the foreword to this book) has been interested in the molecules that help lizards and similar animals regenerate limbs and other tissues. As part of her research she has studied how developmental processes in the species have evolved. For example, a huge family of lizards called skinks has undergone a great deal of evolutionary change. Some have no limbs at all; in others the limbs have become foreshortened, and many have fewer than five toes. Some animals from a genus called *Hemiergis* have only three. In a project carried out with M. D. Shapiro from Harvard University, Rosenthal showed that a long activation of Shh creates more digits; shorter periods create fewer.

CELL DEATH: A SCULPTOR OF DEVELOPMENT

In the early stages of development the limb bud resembles a sort of mitten. Next it goes through a phase in which hands and feet look like the foot of a duck, where early digits are connected by webbing. Later, this tissue "self-destructs" in humans and many other animals to form separate fingers or toes in a process called *apoptosis.* It may seem surprising, but cell death is crucial in shaping many of the body's structures.

Here, too, studies of limbs and digits have been helpful in explaining a more fundamental principle of development. Apoptosis was discovered in studies of limb formation carried out in the early 1970s by Andrew Wyllie, a researcher at the University of Aberdeen, Scotland, and colleagues John Kerr and Alastair Currie. Wyllie was using the electron microscope to compare cells that died because of injuries or toxins to those that died for other reasons. He discovered that injured cells usually swelled and burst open, often leading their neighbors to go through the same process. Cells that died of old age or other causes typically shrank and were quickly taken apart, their fragments absorbed by other cells. Wyllie was the person to give the name *apoptosis* to this second, "natural" type of death.

Scientists had already found cases of developmental processes that required cell death, such as the disappearance of "webbing" between the fingers and toes of human embryos. In the 1960s John Saunders and John Fallon of Marquette University in Wisconsin began investigating this by looking at the feet of ducks and chickens. The feet of ducks are webbed because their toes are connected by tissue, while chickens have separated toes. Their embryos have the webs, though, until a stage at which the cells that make up the connecting tissue die. Saunders and Fallon showed that this self-destruct program could be interrupted if cells were moved at a very early stage.

Experiments in 2000 by Randall Dahn and Fallon showed that cells in chicks depended on signals from the webbing to learn which digit they should form. By removing the webbing, the scientists changed the instructions, making some digits grow in the wrong places. The study showed that a digit learned its identity from the webbing, through signals arriving from the direction of the small toe. Without this tissue it became confused.

The study traced the effects to Shh's control of a transcription factor called Gli3. Normally Gli3 is shut down in cells under the AE ridge, but cells on the thumb side produce Shh and

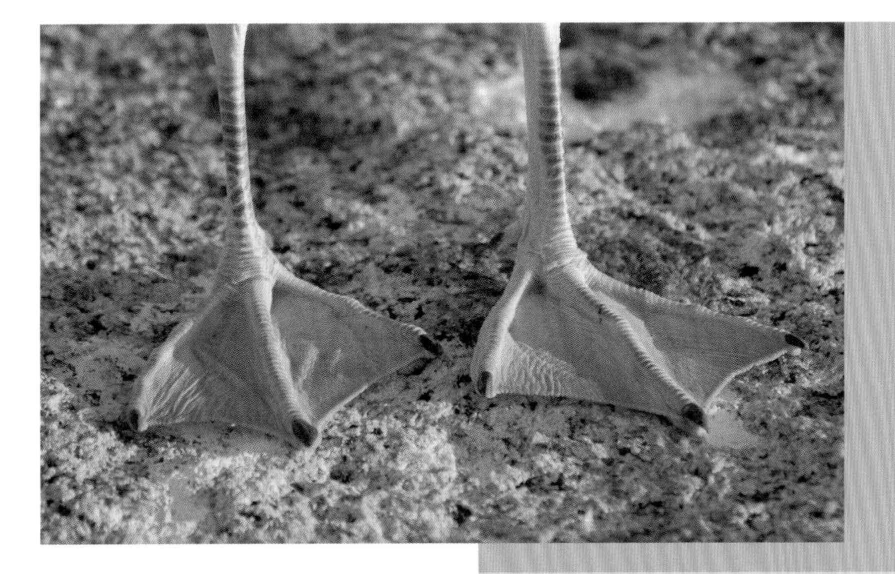

An image showing details of the feet of a herring gull (*Larus argentatus*) *(Eric Baetscher)*

activate a small amount of the protein. Gli3 goes on to help activate Hox genes on this side. To learn more, Ying Litingtung and colleagues in Chin Chiang's laboratory at the Vanderbilt University Medical Center in Tennessee created a strain of mice without Gli3. The animals' legs developed normally except that their paws had too many digits and they were scrambled up, developing in the wrong places.

Apoptosis also separates tissue in the forearm into two separate bones (by carving out a space in the middle) and helps create joints. In the 1990s scientists began to decode many of the molecular signals that trigger these events. Working in Juan Hurle's lab at the University of Cantabria, Spain, Ramón Merino showed that Bmp proteins were probably involved in passing along the signal. The webbing of ducks produces a protein called Gremlin, which blocks it; the cells do not die, and the birds are born with webbed feet. The same pair of molecules is involved in the development of the lungs, described earlier in this chapter. Bmp4 is involved in apoptosis in other tissues as well; for example, it keeps people from growing huge fangs by stopping the growth of teeth after their crowns have formed. It is not, however, a

Chick feet develop without webbing between their toes. This chick is a day old.

universal "self-destruct" command for cells. Bmp molecules were first discovered because they promote the formation of bone, and later they were found to help turn ectoderm into skin.

Apoptosis is now known to act widely throughout the embryo as a means of creating borders and other structures and controlling the growth of tissues. One reason scientists are interested in apoptosis is that it might be used as a tool to fight cancer, as activating the process in tumor cells might make them self-destruct.

EYES

Eyes develop from the same tissue as the brain. They begin as folds on the sides of the upper neural tube, and making them requires the participation of ectoderm, mesoderm, and endoderm in the early embryo. Eyes only form if they receive a signal from the top of the neural plate, a thick region of ectoderm on the back side that will later fold into the neural tube. These interactions produce the "eye field," similar to the limb field discussed earlier in the chapter. Parts of the field receive a signal from the Shh molecule, causing the uppermost region of the neural tube to create two regions that will become the two hemispheres of the brain, the eyes, and other parts of the head.

If Shh is defective or missing, the two sides of the brain become fused and lead to a condition called "holoprosencephaly." In the most extreme cases the embryo only develops one eye, usually below the nose, a rare, fatal problem called "cyclopia." (The name comes from the Cyclops, an ancient Greek mythical figure; some speculate that a real human or animal suffering from cyclopia may have inspired the myth of the giant with one eye.) Shh can be defective due to inherited genetic problems, drugs, or other environmental factors. In the 1950s farmers noticed that sheep were giving birth to lambs with a high rate of cyclopia and other birth defects. Richard Keeler and Wayne Binns of the U.S. Department of Agriculture's Agricultural Research Service traced the source of the problem to the diet of the mother sheep, which were eating a wild plant called the corn lily. The plant contains a potent *teratogen*—a substance that causes developmental defects—that blocks Shh signaling. Cyclopia can also be caused by problems with cholesterol. The reasons are not yet clear, but cholesterol influences how limbs develop by changing the way Shh spreads through tissues (further described below).

If everything goes normally, eye and brain development begins as the top of the neural tube takes on a shape like a brimmed hat. The edges (optic vesicles) grow and curl in on themselves. They come in contact with the outer skin and pass along signals that cause part of it to fold inward. This thickens to create a cluster of cells that eventually detaches from the ectoderm, becoming ring shaped and creating the lens. The remaining cover of ectoderm will form the cornea. Underneath, the optic vesicle begins to produce a protein called Pax6. This molecule seems to play a role in the eyes of every animal species; it appears in all photoreceptors, cells that are capable of transforming light into nerve impulses. An embryo that inherits defective copies of the molecule from both parents does not develop eyes at all, and if the proteins are produced in the wrong place, extra eyes may develop in the wrong tissues.

The cluster of cells that will become the lens signals the ectoderm to develop in a special way to become the cornea. This flexible, transparent external layer has to curve in a precise shape to help focus light onto the retina. The cornea achieves this shape because of the pressure of fluid that collects underneath. The cells that make up the lens undergo unusual development steps that shape them and make them transparent. Initially the lens is a hollow sphere of epithelial cells. Those in the front receive signals that cause them to stretch into long fibers. These grow into the sphere, where they become surrounded by other cells that grow into fiber rings. They produce crystallins, unique, transparent proteins that fill up the cell, eventually pushing out the nucleus. This process takes several years; the crystallins themselves stay intact for an entire lifetime—also very unusual for proteins. While different varieties of crystallins are found in various species, they are all related through evolution, and their ancestors can be found in single-celled organisms that do not have eyes.

If cells in the wrong locations express a gene called "twin of eyeless," ectopic eyes will develop in that area. Here, eyes have grown on the legs of a *Drosophila*. *(Walter Gehring)*

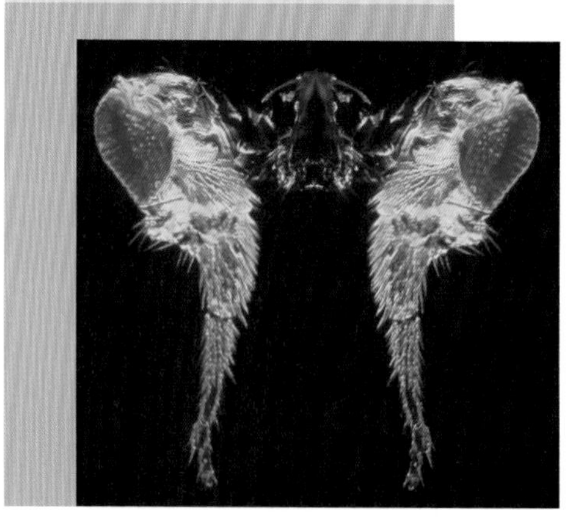

The retina is like a living layer of film. It is very thin—less than half a millimeter thick—but contains 10 layers of distinct types of cells. These are created like a series of waves, starting on the side of the lens and moving inward, with each outer layer passing along signals that guide the differentiation of the next lower layer. Once again Shh signals play an important role. In 2005 developmental biologist Carl Neumann and Ph.D. student Alena Shkumatava at the European Molecular Biology Laboratory in Germany made an important discovery about the signal's function. Cells in the eye—as in the rest of the embryo—reproduce rapidly at first, but this process usually stops when it comes time for them to differentiate into specialized types. In fact, many types of cells show a sort of either-or behavior; if they stop dividing, they automatically begin to differentiate. Shkumatava's experiments with zebrafish showed that Shh signals kicked cells out of the mode of rapid division, and they began taking on the characteristics of eye layers.

Photoreceptor cells stretch through several of the layers, starting with rod and cone "sensors" at the surface. These light-collecting structures are connected to the main body of the cell in a lower level. It extends a wirelike fiber into even lower levels toward the brain. This structure makes photoreceptors both receivers and transmitters.

A protein called *opsin* makes vision possible. It converts light into electrochemical energy that can be passed through the nervous system. Opsin is an extremely old molecule that almost certainly existed in the common ancestor of all life on Earth, which probably lived more than 3.5 billion years ago. Versions of opsin can be found in all branches of life today, from bacteria to all animals and plants. The proteins do not do the same thing in all of these species, but many bacteria and parasites respond to light, and opsins are responsible.

In the eye, opsins sit on the surface of rod and cone cells and are stimulated by photons. This changes the shape and chemistry of the proteins, triggering signals inside the cell that become transformed into nerve impulses. At birth most of these cells are not fully developed, which explains the poor eyesight of babies. By adulthood people have dense photoreceptors that have increased their sensitivity to light hundreds of times.

Rod and cone cells have different structures and functions. Their names come from the shapes of their light sensors. Rods perceive levels of light but not differences in colors; they act like the "brightness" control on a computer monitor. They do not provide very sharp images, but they are especially useful in dim light. Color information is gathered by cones, whose photopigment proteins make them responsive to different wavelengths of light. Most people have genes for three types of these pigments, so they are able to sense red, green, and blue. The input from the different types is mixed by nerves, allowing people to see millions of different shades. Some people lack one or more types of cones, an inherited condition that makes them blind to certain colors. Because the genes for these molecules are located on the X chromosome, men are far more likely than women to be color blind. Men have only one X chromosome; women have two, which gives them two copies of the genes required to make cone cells. If one is defective, the other gene can step in to provide color vision.

In very rare cases some people inherit no cones at all—or the cells are defective—and can only see shades of gray and very strong contrasts. This condition is called "monochromatism." It usually occurs at a rate of about one in 30,000 people, but on the small Pacific island of Pingelap, the rate is much higher: About 10 percent of the inhabitants have no cone cells. This odd situation can be traced back to the late 18th century when a tsunami washed over the low-lying island, killing about 900 of the 1,000 of its inhabitants. Most of the survivors belonged to the royal family. Centuries of inbreeding had spread monochromatism at a high rate through the family. They were likely inside when the tsunami struck, whereas others were outside. The descendants of this family still make up a considerable proportion of the population, and their condition has led to some interesting cultural adaptations. They weave textiles and create other types of art whose patterns only makes sense to people who have monochromatism. The story is recounted in detail in Oliver Sacks's book *The Island of the Colour-Blind.*

While some people lack cones, others have extra genes that permit them to see additional colors. This is known in the animal world: Bees have a fourth photopigment that allows them to see frequencies of ultraviolet light invisible to humans; butterflies have five types. It is hard to imagine how the world appears to these insects—they cannot be asked. In 1993, however, Gabriele Jordan and John Mollon of the University of Cambridge found a person with an additional photopigment. They had predicted that an additional cone would make people sensitive to certain shades of colors that looked identical to most. They carried out an experiment in which subjects were asked to tell whether two shades matched or not. One of their subjects, "Mrs. M," seemed to be a "tetrachromat," with a fourth photopigment allowing her to see an additional color between red and green. Since society and the language used to describe color have adapted around red-green-blue vision, there may be many more people like Mrs. M, who go their whole lives without realizing that there is something different about their vision.

Opsins and other genes that are crucial for eye development—such as the transcription factors Pax6, Six3, and Rx1—are nearly universal in animals and are even found in one-celled organisms that do not have eyes. The opsins often make bacteria and single cells sensitive to light; the other genes have different functions. Their wide presence is powerful evidence for evolution. Like most other animal genes, they arose before cells began to shape themselves into complex bodies. Once they existed, they underwent small changes that allowed them to be used in a variety of ways. Thus, the eyes of flies are radically different from those of humans, but they are built using related genes. The only plausible explanation for this is that they were inherited from a common ancestor, which probably also had some form of eye.

SUMMARY

Following the first steps of gastrulation, embryonic tissues from the ectoderm, mesoderm, and endoderm undergo migrations and growth that bring them into contact with cells at specific locations in the other layers. This causes changes in the genes that they express and triggers the development of new cell types and structures. A relatively small set of structures such as the neural tube and the gut undergo processes of expansion, pinching, folding, and budding to produce a wide variety of recognizable organs. After this stage the organs will continue to grow and undergo refinements that will prepare them for birth.

4

Cell Differentiation and Disease

So far this book has followed the development of an embryo from its origins as germ cells in two parents through the construction of many of the body's major organs. Many of the signaling molecules that direct development have been identified. Scientists still have a long way to go, however, before they have a complete picture of how at least 200 major types of cells arise and take on their proper roles in the body. Understanding these processes will be crucial to finding cures for cancer, Alzheimer's, or cardiac diseases. These problems stem from small changes in cells and genes, some of which occur naturally as a person ages. They have rippling effects that spread beyond single cells and eventually cause major body systems to break down. This chapter examines a few of the main types of cells in the body, discusses their development, and takes a look at their contributions to disease.

A large number of diseases have been linked to developmental processes within cells and organs. Because a book of this size can only address a few, this chapter focuses on three major areas of current research: the link between stem cells, tissue regeneration, and cancer; blood cell differentiation; and finally, a few problems connected to cells in the brain.

STEM CELLS AND POTENTIAL THERAPIES

Nearly all of the body's cells are specialized, but they start as generic stem cells that can develop along several routes. The most generic cell of all is a fertilized egg, which is "totipotent"—it can produce all the cells of the body as well as structures such as the placenta that help the embryo survive in the mother's womb. After just a few rounds of cell replication, however, the daughter cells lose most of this flexibility. There are still stem cells, but they are more restricted. By adulthood most of them are capable of producing only a few types of cells.

As long as a cell is totipotent, it can create an entire human being if it is implanted in a mother's womb. If such a cell splits off from an early embryo, the result may be an identical twin, but cells lose this capacity quickly. Within the first few days of life the embryo has become a ball-shaped blastocyst. The inside of the ball contains embryonic stem cells (ESCs), which are "pluripotent"—able to become nearly all types of cells. They are still nearly totipotent; they can develop into any of the hundreds of types of cells found in the human body, but they cannot create the structures outside the embryo. As the embryo continues to grow, it develops more specialized types of stem cells that are still pluripotent but are only able to generate a limited number of types.

Researchers are particularly interested in ESCs because in many degenerative diseases and processes such as aging, cells die and the body can no longer replace them. Recent experiments in animals have shown that when ESCs are transplanted, they can sometimes develop into mature, healthy cells that can repair damaged tissues. In 2002 Evan Snyder and his lab at Harvard Medical School transplanted ESCs into the brains of old mice suffering from a condition similar to a human stroke or cerebral palsy, in which neurons die. The transplanted cells formed new neurons and reduced some of the symptoms of the disease.

In 2004 Dr. Karen Watters and C. Eduardo Corrales of Harvard and the Eaton-Peabody Laboratory, also in Boston, found that ESCs could grow into cochlear hair cells in the ears of mice. Normally the body cannot replace such cells, and if they die, the result is deafness in both humans and mice. In an experiment carried out two years later Corrales and colleagues Rodrigo Martinez-Mondero and Albert Edge removed cells from an embryonic mouse and implanted them into an adult that had lost crucial connections between the ear and the brain. Within 12 days the cells had developed into neurons, and within about three months they had begun to connect sensory cells in the ear to the brain.

Stem cells in different species of mammals and the signals that cause them to develop are so similar that sometimes the cells of one species can regenerate tissues in another. In 2006 Thomas Reh of the University of Washington and Raymond Lund of the Oregon Health and Science University implanted human ESCs into the eyes of mice that had become blind due to the loss of photoreceptor cells. The procedure restored the animals' vision to 70 percent of that of healthy animals with no eye damage.

In spite of such early successes, research into the possible applications of stem cells in medicine is only beginning. Scientists hope that they can be used to help rebuild severed nerves of the spinal cord as well as brain cells, muscle damaged in injuries or heart attacks, and other tissues. Beforehand, a great deal must be learned about the behavior and possible side effects of transplanted cells. Concerns about how the cells are obtained and used have led many countries to pass laws restricting the kinds of research that can be carried out with them.

The stem cells in adult tissues are no longer as pluripotent as ESCs, so the most useful cells would have to be obtained from an embryo. This raises both ethical and technical issues. As each ESC could potentially become a complete human being, using it to cure a disease might be considered the equivalent of sacrificing one life to save another. On the other hand, the most common sources of these cells are aborted fetuses or embryos created for *in vitro fertilization,* a method that helps infertile couples by bringing egg and sperm together in the test tube, then implanting the embryo in the womb. This procedure requires the creation of extra fertilized eggs that will not be used. In either case, supporters of human stem cell research say it is better to harvest their cells in hopes of saving future lives than to simply destroy them.

In 2007 Robert Lanza of the company Advanced Cell Technology in Massachusetts announced that his laboratory had managed to extract stem cells from eight-cell embryos without harming the donor. Lanza had already achieved a breakthrough as head of the team that cloned the first human embryo; they transplanted the nuclei of such cells into unfertilized eggs in 2001. Some of these eggs developed into blastocysts but were not implanted into mothers to develop further. If the new procedure could be perfected and widely used, it might offer a way to avoid the ethical problems of sacrificing an embryo to benefit another person. Of course, there are other ethical issues. Some people consider that an embryo and its cells should be protected from the moment of conception.

Alongside ethical concerns are scientific reasons why more stem cell therapies have not yet been tried on humans. In some cases the animals produced in similar experiments have developed cancers or other side effects. For example, moving cells from one person to another is the equivalent of a very small organ transplant with a high likelihood that the newcomers will be rejected by the recipient's immune system. This problem might be overcome if a person's own stem cells could be collected for use as later therapies. For this reason some hospitals have started collecting and preserving tissue during births from the umbilical cord; it contains cord blood cells that are virtually as pluripotent as ESCs. Such cells might not work in the treatment of genetic diseases, because a person's stem cells would contain the same faulty information that caused the disease in the first place. This may not be a problem for the treatment of diseases that appear late in life because the new cells might act young, living for years or decades before the problem reemerged. Or it may be possible to treat the cells to correct defects before reimplanting them. This raises more ethical issues, however, because it would involve engineering the human genome, which many consider a taboo. In any case, such treatments have not yet arrived; there are still technical issues to overcome.

Stem cells divide at a fast rate while staying generic. This is not their "default" state. In the test tube they can be kept generic for a long time, but in a tissue they are bombarded by signals that tell them to stop dividing and commit themselves to a course of development. (One process is the flip side of the other; cells that develop usually stop dividing.)

The German researcher Hans Schöler of the Max Planck Institute for Biomedicine in Germany has compared stem cells' development to a marble that rolls down a slope consisting of a series of forking paths: Reaching the bottom means that a cell has become fully specialized. Just as a brake or some other device is needed to keep the marble from rolling, Schöler has helped show that stem cells have to produce a molecule called Oct4 to keep their generic status. Without the protein the cell starts to move along a developmental pathway.

The road between a fertilized, totipotent egg and the hundreds of types of cells in a human body has many forks, some of which are discussed in the following sections. As an example, the diagram on p. 118 shows the steps required to make the many types of blood cells found in humans. Each fork represents a step that commits a cell to some fates and prohibits others. The decision is usually triggered by a particular transcription factor. Often the cells get put on hold along the way. They take a few steps toward specialization but stop as "adult" stem cells that may wait for long periods of time before going the rest of the way. This gives the body reserves of cells that can be called up to repair damage. Bone marrow carries rich stocks of the cells that develop into blood cells; satellite cells at the edges of muscle fibers can be used to replace damaged cells. But many tissues do not have such reserves, so the body cannot repair major damage to the brain, heart, or spinal cord, for example.

In most tissues of humans and mammals, once a decision has been made, there is normally no natural way of going back, just as it would take something very unusual to make a marble roll back uphill. (In some other animals, such as newts, this is not the case.) But some researchers believe it might be possible to take partially committed cells and make them more pluripotent, which might allow scientists to avoid having to use ESC. Achieving this would require reversing the effects of transcription factors and reactivating the programs that hold cells in their generic states.

Few people object to work that is being done with adult stem cells, but these are often difficult to find in the body and grow in the laboratory, and they may not be able to integrate themselves into tissues as well as embryonic cells. In 2006 Eugen Kolossov of the German company Axiogenesis and Bernd Fleischmann's laboratory at the University of Bonn looked at

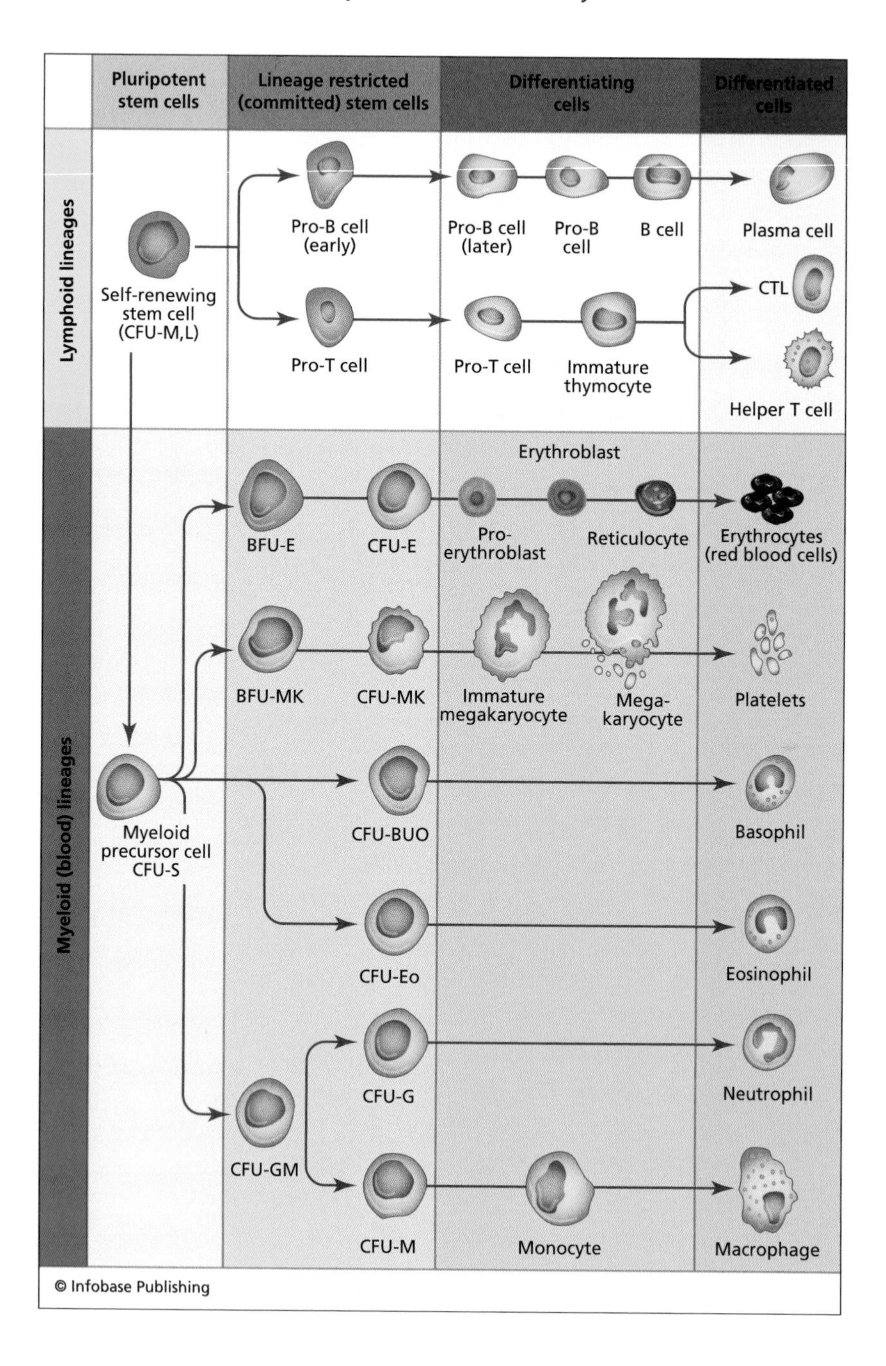

the capacity of stem cells to repair damage in mice that had suffered heart attacks. Transplantations of bone marrow, which contains adult stem cells, had no effect, but significant improvements in the hearts were seen when the scientists implanted ESCs that had been cultured in the lab and transformed into cardiomyocytes, a type of stem cell that develops into heart muscle.

Very recently there have been a number of breakthroughs in research that open up new ways of working with stem cells, including adult types, and possibly new therapeutic applications. The work involves *induced pluripotent stem cells* (iPSCs), which are made from cells that have specialized. Usually the process of differentiation is considered a one-way road—cells do not reverse the steps they have taken to become more generic again—but with the discovery of genes that help maintain stem cells, researchers began to wonder whether this was really true. What would happen if a differentiated cell were forced to use stem cell genes?

In 2006 Shinya Yamanaka's laboratory at Kyoto University in Japan tried the experiment with mouse cells called "fibroblasts." These cells provide a structural framework for tissues throughout the body and participate in wound-healing processes. Yamanaka and his colleagues used viruses to transfer four genes to the cells, which then began to exhibit stem cell properties. When they were added to early-stage mouse embryos, the cells failed to act like ESCs. But, with a slight change in methodology, by 2007 the Japanese group, along with labs at the Massachusetts Institute of Technology (MIT) and the University of California at Los Angeles, succeeded. The cells were integrated into the embryos and were used to produce a wide range of body tissues. Later that year Yamanaka's lab and another group headed by James Thomson at the University of Madison–Wisconsin managed to turn human fibroblasts into healthy iPSCs.

(opposite) Hematopoeitic stem cells go through several stages of development on their way to becoming over a dozen types of fully differentiated blood and lymph cells.

Illuminating the Cell

In 1962 Osamu Shimomura, working at the University of Washington's Friday Harbor Laboratories, purified a green fluorescent protein (GFP) from a jellyfish; the protein was responsible for the creature's beautiful fluorescent colors. The discovery remained an interesting curiosity until the early 1990s, when Douglas Prasher of the Woods Hole Oceanographic Institution in Massachusetts cloned the gene. (Cloning shows scientists exactly where a gene is and what it is made of, allowing them to make a huge number of copies of the molecule—a first step in investigating how it works and using it in experiments.) Martin Chalfie, a researcher from Columbia University in New York City, heard about the molecule at a conference and immediately realized that it might be turned into a very powerful tool for biologists.

His laboratory inserted the gene into bacteria and discovered that species other than the jellyfish could use it to build a fluorescent protein. The next step was to use GFP as a tag added onto other genes. Scientists began tagging other molecules in mice and other organisms. Usually this led the animal's cells to produce a form of the protein that would light up when examined under a laser microscope. It gave researchers a new tracking device to watch molecules as they were made and moved through the cell. This revealed, for example, where and when a protein became active in an organism. This technique has been used to make a map of the developing eye, with several colors, showing a different molecule typical of a different type of cell. The fluorescent tags show how signals move in wavelike patterns through layers of cells, starting at the forward edge and moving back through successive layers of cells, triggering their differentiation into all the types needed to make up the eye.

Before such images could be made it was necessary to brighten the color of GFP—the natural jellyfish molecule does not radiate much light—and add new colors. Some of these have been taken from other organisms, and all of them have been enhanced through genetic engineering. Mutations have been introduced into the fluorescent molecules, intensifying them or changing their shade. Now scientists can draw on a wide palette of fluorescent proteins. This permits them to tag multiple genes in the same cell and study how proteins interact or influence one another's behavior.

GFP and the newer tags have triggered a revolution in light microscopy. The physics of fluorescent molecules permit entirely new types of experiments. The proteins emit light because they absorb energy from the laser and then give it off again. Each protein only does this one time, so if the laser focuses on one area of the cell, it eventually "bleaches" all the fluorescent molecules there. But the cell may make new proteins and move them into the bleached area. Watching this effect over time has allowed scientists to measure how fast proteins wear out and how quickly the cell replaces them. Another type of experiment involves measuring how fast the energy that has been absorbed is released again. Molecules do this at different rates depending on their shapes and their activity. This energy discharge can be measured very precisely, so it can give scientists an idea of what a particular protein is doing.

Another kind of experiment, called fluorescence resonance energy transfer (FRET), involves what happens to the energy after it is released. If two different fluorescent molecules are docked onto each other, they absorb some of each other's energy. This slightly changes the amount that they emit. That can also be measured, giving scientists the first method of determining whether two proteins are directly in contact. One common use of FRET is to study how signals arrive at the cell and how they are passed along to the nucleus.

In 2008 Samuel Wood, a researcher at the California biotech company Stemagen, successfully used his own skin cells to produce iPSCs and clones that survived five days.

One day, scientists hope, it will be possible to routinely take fibroblasts or other differentiated cells and make them totipotent again, or at least pluripotent. This could allow researchers to take cells from an adult and use them to regenerate particular cell types or tissues that have been damaged in a disease or an injury. It then might not be necessary to harvest ESCs. But many years of research and testing will be required to learn whether iPSCs are really as potent and useful as ESCs and to ensure that their implantation in an animal or human does not cause new types of health problems.

STEM CELLS AND CANCER

The modern understanding of cancer dates back to the mid-19th century, when the German physician Rudolf Virchow (1821–1902) proposed the theory *Omnis cellula e cellula*: Every cell arises from a similar cell that already exists. Virchow quickly saw the connection to cancer: Tumors, too, originate with a cell—a diseased one. With the discovery in the mid-20th century of the role that genes play in producing healthy cells, researchers made another crucial step toward understanding the nature of the disease. Cancers arise, they believe, from one or more mutations affecting cell division and specialization. The problems cause cells to get stuck somewhere along the path of differentiation. They should no longer divide, but for some reason they do. The process gets out of control and evades bodily defenses that normally kill defective cells. In a further stage of the disease the cancer develops a *metastasis*: Cells migrate out of the original tumor, settle somewhere else, and grow. Eventually this disturbs the function of an important organ. While solid tumors can often be removed by surgery, metastases are often hard to find until it is too late, which is why they cause 90 percent of deaths linked to "solid" tumors.

One link between stem cells and cancer is the idea that the cause lies with defects in stem cells themselves. Another is the idea of a "cancer stem cell," which has been proposed as scientists try to understand the origins of metastases. Most researchers have believed that the process begins when cells in the tumor undergo additional mutations. But, according to a new hypothesis, even in very early stages some tumors may contain a sort of stem cell that has the potential to metastasize.

The "textbook model" of a series of mutations is probably true in many cases. Colorectal cancer (CRC) is an example. Under the microscope, the inner lining of the intestines (which draw nutrients from food) is interrupted by a large number of deep, well-like holes called "crypts." At the bottom of each crypt is a population of stem cells that reproduce, then climb out to rebuild the intestinal lining as its cells wear out. CRC frequently begins when there is a mutation in the pathway that interprets a signal called Wnt (introduced in the first chapter). Wnt tells stem cells to divide. The laboratory of Hans Clevers at the Netherlands Institute for Developmental Biology showed that if the signal becomes too active, cells can start to form tumors, but the tumors quickly die unless they escape the crypts. Eduard Battle and Elena Sancho, two molecular biologists in Clevers's group, have shown that an escape requires changes in the signaling system triggered by two other molecule types called Ephs and ephrins.

The advent of new technologies such as DNA chips (described in chapter 3) has given scientists new ways to investigate tumors and look for the factors that cause metastases. One method is to scan the gene activity of secondary tumors, created when metastatic cells spawn a colony in a second tissue, and compare it to that of cells that have remained in the original tumor. This often reveals differences in the way the cancer cells use particular genes, which can be used to find a "signature" for the metastatic cells. Culprit genes thought to cause the metastasis can be tested by activating them in cancer cells that do not have them and watching to see if they become metastatic.

In some cases the reasons for the differences can be traced back to mutations, but sometimes the signature can already be detected in cells of the primary tumor. This hints that it may already contain cells with the potential to become metastatic, or cancer stem cells. This may even be true of some CRCs. In 2008 Ulrike Stein and Peter Schlag of the Max Delbrück Center for Molecular Medicine in Berlin discovered such a marker. They found that cells that produced a protein called MACC1 were likely to develop into metastases. If traces of the protein could be detected in the primary tumor, there was a very high probability that the patient would develop metastases. A test for the signature is being developed into a diagnostic tool to help scientists plan treatments for their patients.

Many other connections have been found between tumors and stem cells. Mary Hendrix and her colleagues at Northwestern University in Chicago have found that the protein nodal helps metastatic cells communicate with tissues they have begun to invade. Nodal was introduced in chapter 3 for its role in embryonic development; it is involved in keeping stem cells on hold. In 2007 the researchers showed that by shutting down nodal in skin cancer cells in zebrafish and chicks, they could greatly reduce the cells' ability to spread.

Another feature of many tumor cells is their unusually long lifespans. Most cells go through a limited number of divisions before specializing or dying, partly because of the way DNA is copied. The DNA in most animal cells is arranged in sticklike chromosomes. The machinery that copies it cannot reach all the way to the ends, so each round of division usually leads to shorter and shorter chromosomes. Normally, losing DNA would have bad consequences, but evolution has produced two solutions. The ends of the chromosomes contain structures called *telomeres,* long stretches of meaningless information that can be cut off without causing any damage. This can only go so far, though, before the organism begins to lose genes. Here the second device comes in: Molecular machines called *telomerases* add extra DNA to the ends of the chromosomes, partially making up for what has been lost. Telomerases were discovered in 1984 by Elizabeth Blackburn, an Australian-born professor of biology at the University of California at Berkeley, and a Ph.D. student named Carol Greider.

Both stem cells and cancers activate telomerases, which wins time before damage sets in to stop the cells from reproducing. Other mechanisms keep them from differentiating. In 2007 Maarten van Lohuizen, a researcher at the Netherlands Cancer Institute, found that a group of molecules called "polycomb proteins" might be responsible for this developmental block in cancer cells. Polycombs place chemical tags on DNA that usually prevent genes from being activated. Lohuizen and his colleagues created strains of mice with defective forms of the proteins. Tissues in the animals divided too often and released cells that migrated and grew elsewhere, much like cancer cells do.

There are several other parallels between tumors and stem cells. Stem cell transplants sometimes lead to tumors in the new host. The reasons are not clear: The relocated cells might be confused by signals from their new neighbors, or the opposite might be happening—the transplants may be stimulating other cells to behave in the wrong way.

Cancer cells do other things in adults that resemble embryonic development: They push the formation of new blood vessels, probably by triggering the development of nearby stem cells. They use embryonic signaling molecules to stay in a "copying" mode. Mutations in important developmental molecules like Shh, Wnt, and Notch (introduced in the previous chapter) often give rise to cancer. Leukemia is thought to start when hematopoeitic stem cells (HSCs), which give rise to blood, undergo mutations and do not develop properly. Blood cell development is discussed in the next section.

The differences between embryonic and other stem cells and tumors will have to be studied much more intensively before stem cell therapies can be tried safely in humans. So far, experiments have been restricted to animals or to human cells that were destined for destruction anyway. This does not solve all the ethical issues regarding work with stem cells, but blocking such work also raises a significant ethical problem. Cancers and degenerative diseases cause an immense amount of suffering. No one can predict how well or how soon stem cell research might lead to treatments, but it will certainly happen much more quickly if scientists are allowed to work with human cells in a careful and reasonable way.

BLOOD CELLS

Nine major types of cells arise from HSCs. These types become red and white blood cells as well as the components of the lymphatic system, a second type of circulation that develops alongside blood vessels as a route for the transport of fats and fluids. Lymphs also play a crucial role in the immune system by filtering invading microorganisms from the blood. This section introduces the development of a few of the types of cells and their connection to disease.

The HSC "factory" moves during embryonic development. The first stem cells for blood appear in clusters of cells called blood islands, which also produce vessels (introduced in chapter 3). Later, the center of production moves to a position near the aorta, one of the major vessels of the heart. By adulthood the stocks of partly developed stem cells have relocated to the bone marrow. A transcription factor called SCL directs more generic cells to become HSCs, and other proteins push them to become the more specialized types.

Red blood cells (erythrocytes) are born in the marrow of the body's major bones. They are stimulated to grow by a hormone called erythropoeitin, produced in the kidney. Over about seven days the cell undergoes some strange transformations, particularly in humans and salamanders. In these two species the cell ejects its nucleus and some of its other organelles, leaving it doughnut shaped, with a depressed area rather than a hole. This shape makes the cell more efficient at its major task of taking up oxygen and transporting it through the body. At the same time, it limits the cell's lifespan; without the nucleus the genes are gone and cannot be used to replace RNAs and proteins. Erythrocytes survive for about 120 days until being taken apart by another type of blood cell, a phagocyte. The short lifespan means they have to be constantly replaced, at a rate of about 2 million new cells per second, by calling up stem cells from the bone marrow.

Red blood cells are able to carry oxygen because they host vast quantities of a protein called hemoglobin. A single cell holds more than a quarter million copies of the protein. Each hemoglobin copy holds four atoms of iron that become scarlet red when loaded with oxygen, giving the cell its color. Hemoglobin has been the subject of intense research for more than 60 years not only because of its importance to the body but also because it is linked to several very serious diseases, including sickle-cell anemia.

Some mutations in the hemoglobin molecule disturb the development of red blood cells. Instead of the doughnut shape they take on a form like a crescent moon, or a sickle. Inheriting two copies of such a mutation, one from each parent, is fatal. People who inherit only one copy often have serious health problems because their blood transports less oxygen, and the strange shape of the cells leads to strokes and other types of damage. Normally, evolution would surely weed out this type of mutation, but sometimes it can be helpful to its carrier. Many regions of the world suffer from high rates of malaria, caused by a mosquito-borne parasite. This organism, *Plasmodium,* enters the body and reproduces in red blood cells, but it cannot gain a foothold in sickle-shaped cells. So, in areas infested by malaria, people who inherit one defective copy of hemoglobin sometimes have an advantage at survival; they pass along their genes more frequently, and the mutation is found in the population at rates much higher than in other regions. This is strong evidence that human evolution has been partly shaped by parasites.

Several other diseases have been linked to mutations that affect red blood cells. Hemoglobin requires iron to function; if the body lacks it, the molecule will not be loaded with oxygen. Most of the body's stock of iron comes from recycling, supplemented by very small amounts absorbed from food, but in some genetic diseases one of these processes is disrupted. Diseases called "erythrocytoses" occur when the body produces too many red blood cells compared to other types. Because they are some of the smallest cells in the bloodstream, this causes the blood to flow too quickly.

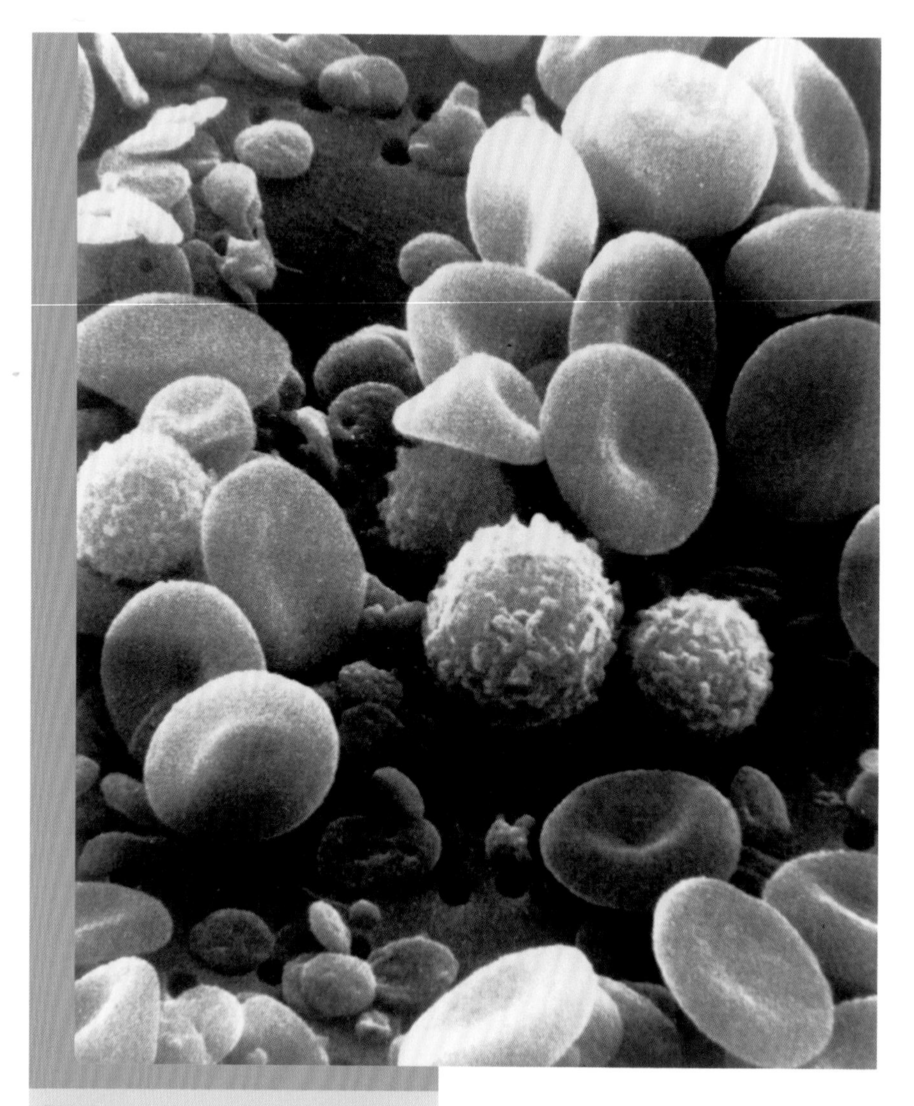

This image of normal circulating human blood was produced by a scanning electron microscope. In it one can see red blood cells, several white blood cells including lymphocytes, a monocyte, a neutrophil, and many small disc-shaped platelets. *(National Cancer Institute)*

In addition, HSCs produce several types of white blood cells (leukocytes). These act as watchdogs of the immune system, traveling through the bloodstream and slipping through the walls of vessels to track down bacteria, viruses, and parasites. They have different strategies for coping with invaders. Some types engulf the foreign particle and digest it with enzymes. Others approach it and burst, releasing highly reactive substances such as oxygen that cause the invader to break down.

Neutrophils are the most common type of white blood cell and serve as first-aid teams, rushing in at the first sign of a bacterial infection. They sweep up debris and carry it away into the lymph system for disposal. Monocytes are similar but also have a role in training other types of cells to recognize infectious agents. If they leave the bloodstream and enter the surrounding tissue in search of invaders, they undergo additional development and become macrophages. These are the large scavengers of the immune system—made famous at the close of the film *Fantastic Voyage,* where the cells digest a miniaturized submarine.

Insects have a simple immune system that responds to infections by trying to destroy anything that looks unfamiliar. Humans and other mammals have evolved more sophisticated responses, thanks to blood cells that teach one another to recognize specific viruses, bacteria, and sometimes even cancer cells. This is achieved by B and T cells, which develop in the lymph system.

B cells are constantly being made from stem cells in the bone marrow. They produce surface proteins called "antibodies" that may be able to bind to a foreign protein on an invader. The body cannot guess ahead of time what such molecules might look like, but B cells have a special mechanism that rearranges antibody genes in random ways, like making randomly shaped puzzle pieces in hopes of fitting one that has fallen out of the box. The process allows the body to create more than 1 trillion different antibodies, each with a unique conformation that is able to recognize a different potential partner.

The B cell circulates through the blood and lymph system on the lookout for parasites or toxins. If its antibody matches a foreign molecule (an *antigen*) and then comes in contact with a helper cell, it may undergo another developmental step. It becomes a large plasma cell, a factory to turn out huge amounts of copies of the antibody. When released, these molecules can bind to antigens on other copies of the invader, marking it for destruction. Other B cells turn into "memory" cells that live for a long time and can launch an immune response instantly if the parasite tries to infect the body a second time. This is the reason that many vaccines need to be given only once during a person's lifetime or repeated only after long intervals.

If antibodies are made at random, how do they avoid matching the body's own proteins and calling for the destruction of its own cells? The answer lies with a small group of HSCs that move to the thymus, a gland located at the base of the neck. Here they develop into more types of white blood cells called "T cells." Before they are released from the thymus, T cells are put through rigorous tests that teach them the difference between native and foreign molecules. One of their later tasks will be to bind to molecules that are attached to a protein called the *major histocompatibility complex,* or MHC. The aim of the first test is to make sure they can do this strongly enough. The T cells that fail are destroyed. The survivors then move on to a region of the thymus called the "medulla," where they undergo a second test. Its goal is to weed out T cells that dock too strongly onto the body's own proteins. Only about 2 percent of the cells pass both tests. Most of those that survive are released into the bloodstream. A few become regulatory T cells, which stay behind to act as a sort of police force for the immune system.

Some of those that are released become helper T cells. They make contact with B cells that have found an invader and then sound a general alarm for the immune system. The cells divide quickly and release small proteins called cytokines, which amplify the immune response. The mark of a helper T cell is that it produces a protein called CD4, which is important for an entirely different reason. HIV (human immunodeficiency virus), which causes AIDS (acquired immunodeficiency syndrome), recognizes CD4 and uses it as a passport to enter the T cell. It may take several years for AIDS to break out in the body; if that happens, the first victims are the infected T cells. Their loss causes the breakdown of the immune system and the symptoms of AIDS, in which other infections ravage the body.

A second type of T cell produces a protein called CD8. These cytotoxic T cells roam through the body, searching for cells that have been infected by viruses and destroying them. This type also bears most of the responsibility for rejecting organs after transplants.

Toward the end of an infection the immune system is usually running in such high gear that it begins to see enemies everywhere. Because there is a danger that it will start attacking healthy cells, the process has to be shut down, and this is the job of regulatory T cells. By weakening the response of other cells to infections, they protect the body from overzealous immune responses. They have a second job in the thymus, where they weed out other types of T cells that bind too strongly to the body's own proteins. If these cells do not function, or if the body does not produce enough of them, the result will be a severe autoimmune disease. A protein called Foxp3 seems to play a key role in their development and functions. People with mutations in Foxp3 develop a disease called IPEX, in which the immune system cannot be shut down and spins out of control. Other autoimmune problems that have been linked to the regulators are type 1 diabetes, multiple sclerosis, and several types of arthritis.

Regulatory T cells therefore walk a fine line between assisting and hindering the immune system. The fact that they target misbehaving cells has led scientists to wonder whether they might be "reprogrammed" to destroy tumor cells. This would require outfitting them with receptors that recognize markers that appear only on cancer cells. The strategy has already been used to fight diseases in animals. In 1994 Youhai Chen and Howard Weiner's lab at Harvard Medical School showed that the cells could be "trained" to combat an autoimmune disease in mice called experimental allergic encephalomyelitis.

Most types of leukemia arise from problems in the development of other types of white blood cells. Stem cells in the bone marrow divide too rapidly and produce an overwhelming number of immature cells that do not usually function, leaving the body open to infections. They overpopulate the marrow and interrupt the development of other cells, such as red blood cells. Eventually the body is not supplied with enough oxygen, organs are damaged, and other severe health problems arise.

NEURONS AND GLIA

A palm tree has a bushy crown, a long, slender trunk, and a huge, sprawling network of underground roots—much like neurons in the human brain. These cells have rootlike dendrites, branching extensions that reach out in all directions to receive information from neighbors, putting a single cell into contact with as many as a thousand others. They spread out close to the main body of the cell, called the "soma," which contains the nucleus and other cellular structures. Additionally, each neuron has one axon, like the tree's trunk and crown. It may be tens of thousands of times as long as the soma.

Axons are broadcasters, releasing small molecules called "neurotransmitters" that are captured and absorbed by the dendrites of other cells. This happens at an amazingly fast rate: The speed at which a signal moves across the brain can be up to almost 7.5 miles (12 km) per second. Substances like caffeine increase this speed; other substances slow it down.

Neurons arise from cells in the neural tube, and the brain eventually contains about 100 billion of them. Even though it sounds like a huge number, neurons are a minority in the brain. There are 10 times as many glia, which develop from the same type of stem cell. These numbers may be the source of the (completely untrue) myth that people "only use 10 percent of their brains." Glial cells have many functions: They act as scaffolds for neurons, feed them, and also participate in communication. But they have received much less attention than neurons, so scientists know much less about them.

In early stages of development the body makes neurons at an amazing rate—up to 250,000 cells per minute. This pace has to slow down, and in a person's teen years the cells are "pruned" either through the death of neurons or the paring down of some of their connections. These events partly explain the way people acquire particular mental abilities at some points in their lives and lose them at others. For example, if someone is intensively exposed to a language before puberty, he or she can usually master it to the level of a native speaker. Later that becomes impossible for most people; even with an immense amount of

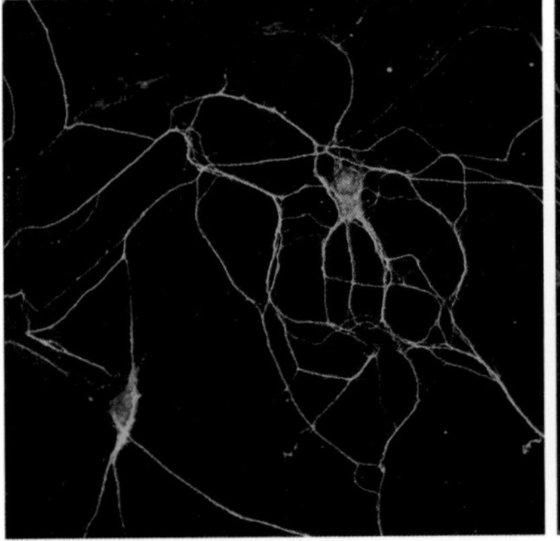

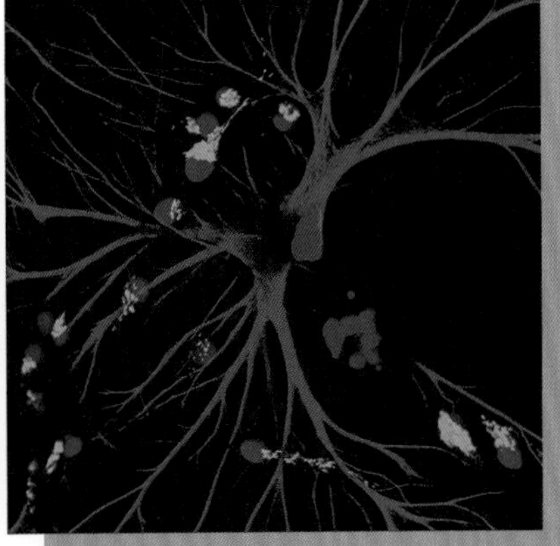

Neurons and glial cells *(University of Bath, Centre for Optical Studies)*

practice in a foreign language, they will be unable to lose their accents and master all of its nuances. This probably has to do with changes in the connections between cells as well as the function of the brain as a whole. After a certain age there is a shift in the regions used to process a new language.

In the past, scientists believed that the body stopped creating neurons in very early childhood, but recent advances in brain imaging techniques have changed that view. Jay Giedd and Judith Rapoport of the National Institute of Mental Health in Maryland have been using magnetic resonance imaging (MRI) to study normal and abnormal brain development in children and young people. This method gives scientists a three-dimensional view of the brain and how its regions are activated as test subjects carry out different types of activities.

The researchers have shown that various parts of the brain develop at different rates and have found some clear differences between normal and abnormal development. Recent interesting projects include studies of twins with autism, schizophrenia, and hyperactivity, trying to distinguish the roles that genes and environmental conditions play in these disorders. Over the course of their work Giedd and Rapoport have revised the traditional view of neuron development. One discovery is that cells continue to develop until adolescence—longer than previously thought. Then they stop in most parts of the brain, except for a few places, including the hippocampus. This region of the brain is crucial for memory, learning, and navigation.

Glial cells continue to be made throughout a person's lifetime, and a study carried out in 2006 by Dennis Steindler's laboratory at the University of Florida showed that the cells may even have more functions than previously thought. The scientists removed glia from human patients suffering from epilepsy and cultured them in the test tube. When they added proteins that stimulate cell growth, the glia took on characteristics of stem cells and began to divide. Transplanted into a mouse, they seemed to develop into neurons and other types of brain cells. The results mean that glia can be reprogrammed—at least in the laboratory—to become a sort of pluripotent stem cell. The discovery might lead to therapies in which glia can be used to replace damaged neurons or other types of brain cells.

The nervous system only functions if its cells take on the right shape and structure. This is a staggering task; a type of cell called a "Purkinje neuron" may form more than 100,000 connections to other cells by growing massive numbers of dendrites. An important question is why the cell sprouts so many dendrites but only one axon. The problem is especially interesting because in the early stages of differentiation both structures look the same, forming from identical small spikes protruding from the cell surface. In 1999 Frank Bradke and Carlos Dotti at the European Molecular Biology Laboratory in Heidelberg set out to find out what makes the axon special. They found at least a partial answer having to do with the interplay between microtubules and protein fibers that give the cell its shape and structure.

The first chapter introduced microtubules: long, pipelike assemblies of molecules that act as both scaffolds and a railway through the cell. Microtubules form the backbone of the axon as it extends for huge lengths through the body and also the tracks for molecules that shuttle back and forth to the extremities. The cell's second major scaffold, a weblike network made of proteins called "actin," also has an important job in the axon. Its tip ends in a bushy region called the growth cone, thousands of fingerlike actin fibers reaching out in all directions to "taste" molecules on nearby cells. Sometimes the taste is repellent and a finger withdraws; at other times the signal tells the axon where to go. The protein laminin marks the correct route, and ephrin (introduced earlier in connection to colorectal cancer) usually acts as a wrong-way sign. When it arrives at its ultimate route, actin fibers help glue it to target cells so that it does not wander off again.

The developing neuron has a dense zone of actin at the outer rim, just inside the membrane. These fibers block the path for microtubules, but Dotti's laboratory discovered that a pathway is cleared at one point, allowing the microtubules to penetrate and push at the cell membrane. This is the place where the axon develops. By using a drug to break down actin in several places, the scientists could create cells with more than one axon. Normally the cell prevents this from happening.

Once the axon has formed, it has to grow in the right direction. Human perception, behavior, and consciousness depend on the axon of each cell finding its way to precise targets. It has to be "hard wired" to the dendrites of specific cells so that sensory information, for example, can be routed to the part of the brain that manages it. The vision centers are located at the back of the brain, so nerves that capture light in the retina have to extend their axons more than 8 cm (a little more than 3 inches)—an incredible distance at the scale of a cell when considering that they have to hit a target that may only be about 1 mm (a fraction of an inch) in size.

Still, this feat of pathfinding pales in comparison to the axons that give the brain control over the legs; they grow to a length of more than a meter (several feet). When the destination has been reached, the wandering axon and its target start a dialogue in the chemical language of neurotransmitters. These tiny molecules are released by axons into gaps between the cells called "synapses." Usually this causes a change in the behavior of the neuron's membrane. Tiny pores are opened or closed, allowing charged atoms to pass through. The result is that signals are amplified or tuned down.

By puberty the features and main "wiring" of the brain are in place, but the organ has not yet finished developing. Experiences and activity change the details of its structure. Learning and memory somehow have to physically alter the brain, and in 1966 two scientists began to understand how. Timothy Bliss and Terje Lømo, students in the laboratory of Per Andersen at the University of Oslo, Norway, had implanted microelectrodes into the brains of anesthetized rabbits to study neuron signaling in the hippocampus. They discovered that if they stimulated the same pathway of synapses again and again, neurons would begin to amplify the signal, similar to turning up the volume of a radio when a favorite song comes along. They also noticed a "use it or lose it" response: Synaptic connections that were rarely stimulated weakened and sometimes even disappeared. Bliss and Lømo called the phenomenon "long-term potentiation" (LTP) and believe it is one of the foundations of memory and learning.

Neurons need the support of glia to survive; they also need contact with their neighbors. In Alzheimer's and some other neurodegenerative diseases, fragments of a protein accumulate between cells and form clumps that cannot be dissolved. Eventually this interrupts communications between parts of the brain and to the rest of the body, resulting in death. For Alzheimer's disease the culprit is usually a protein called APP. Despite the fact that this molecule has been intensively studied, its normal functions in the healthy brain are still unclear. Recent work with strains of mice that lack the molecule show, however, that it plays a role in transforming stem cells into neurons, and it also participates in memory and learning.

The problems begin when a portion of the molecule is sliced off and released into the space between cells. There it binds to other fragments and creates fibers and clumps that cannot dissolve. Eventually this interferes with the activity of neurons and other cells. Other proteins are known to aggregate and have similar effects on brain cells, so there are probably many ways that Alzheimer-like conditions develop. Over a lifetime it is probably normal for protein fragments to accumulate; degenerative diseases seem to be a natural side effect of aging. This does not mean that they are incurable, but it makes the search for treatments much harder.

SUMMARY

The body contains more than 200 types of cells that arise from more basic stem cells. As researchers decipher the molecular signals and genes that guide their differentiation, they have discovered numerous links to cancer, Alzheimer's, and a number of other diseases. Today the major causes of death in the developed world are illnesses such as heart disease, cancer, and neurodegeneration. These diseases are usually caused by malfunctions that naturally occur within cells as the body ages. Evolution has not given the body natural defenses against them, because they normally arise after the reproductive phase of a person's life. So learning to treat them will require a thorough understanding of cell differentiation and of how things break down over time, as well as new methods to intervene in basic cellular processes.

5

Evolution and Development

Visitors to the University of Padua in northeastern Italy can see a massive wooden lectern, complete with stairs, from which Galileo Galilei, the famous astronomer, delivered geometry lectures to students. In a nearby building they can visit a tiny round amphitheater that was built during Galileo's days for the medical school. Students stood on circular ramps overlooking a table in the middle, where Girolamo Fabrici (1537–1619) dissected the bodies of criminals who had been put to death. An Italian judge had recently given the school permission to do dissections on human beings, which had always been considered taboo for religious reasons. The practice was still considered controversial, and doctors feared visits from the police, so the dissection table had an ingenious design. A dead animal was mounted upside down on the underside while a human corpse lay on top. The table could be flipped over in an instant, and unwelcome guests would walk in on an uncontroversial dissection of an animal.

The Italians quickly discovered that the classic books on human anatomy, written by the ancient Greeks and Romans, were based on dissections of animals rather than people. Ethical concerns have always led doctors to use animals in place of humans in biological and medical studies. But how similar were they to people, and how much could really be learned from them? No one knew what to make of the fact that the bodies of humans and many animals were composed of mostly the same parts, arranged in a similar way. With the invention of powerful microscopes, scientists discovered that the resemblance reached down to the level of cells. And in the age of genomes, scientists have discovered that humans share virtually their entire genetic code with other species. The explanation, of course, lies with evolution. Charles Darwin's theory allowed the grouping of wildly different species into one family tree and revealed that similar features could be traced to common ancestors. They likely arose through common developmental processes, but evolution said nothing about the mechanisms by which this happened.

Since about the 1980s that situation has changed dramatically with the unification of fields in a new type of science called evolutionary developmental biology (informally, evo-devo), which is the subject of this chapter. Its approach is to compare the activity of genes in many species in order to understand how they cause similarities and differences in development. Along the way researchers have shed light on questions such as the ancient origins of the eye and the brain. The strategy has also allowed them to create sharp pictures of some of the earliest animals that ever lived—even though no fossils of these creatures remain. This work has very practical results: It is allowing scientists to create new animal models of human diseases both to understand why defective genes disrupt the course of development and to try out new forms of therapy in animals before working with human patients.

ERNST HAECKEL AND RECAPITULATION THEORY

In the early 19th century anatomists such as the Frenchman Georges Cuvier (1769–1832) noticed some unexpected and fascinating parallels between the bodies of different species. The fins of porpoises and whales bore almost no resemblance to those of fish, but the bones inside matched those of the human hand and forearm almost one to one, as if the parts of a hand had been stretched and reshaped. The hinged bones around the jaws of reptiles corresponded to the bones that made up a mammal's ear. Decades later evolution would offer answers; for example, porpoises and whales were mammals that had moved from land to the sea, and the ear had evolved from the bones and other structures that formed the face.

Karl Ernst von Baer (1792–1876) noticed something similar in his studies of embryos: Although chickens, dogs, and human beings were strikingly different as adults, they looked nearly identical during the early stages of development. Here, too, evolution would offer an explanation. Most of Darwin's work deals with the adults of species, but he knew that every point in an organism's lifetime might be crucial to its survival. So it was also necessary to look at animals at different phases of their lifetimes.

Along with their genes, organisms inherit the processes by which cells develop into bodies. This principle helped Darwin prove that the barnacle was an arthropod (as were insects, spiders, and shrimp) because of features that appeared during its larval stage. In 1866 Alexander Kowalesky showed that the sea squirt, a baglike marine animal, was related to the chordates (vertebrates and their close relatives). It had previously been classified as a type of mollusk, but Kowalesky observed that its organs arose from the same layers as chordates, which proved the relationship beyond any doubt.

The most famous comparative embryologist would be one of Darwin's most enthusiastic followers, Ernst Haeckel (1834–1919). Haeckel was born in Potsdam, Germany, near Berlin, and received a degree in medicine before deciding that he was cut out more for a life of research than one of dealing with sick patients. He moved to the University of Jena, where he obtained a doctoral degree in zoology and stayed there as professor of comparative anatomy until he retired. He was such an enthusiast—and a prolific writer—that in one meeting with Darwin, he presented him with a 1,000-page manuscript. Darwin did not speak German, and the only way he could deal with the book was to look up every word in a dictionary. (At some point he became exhausted and gave up.)

Haeckel followed in the footsteps of von Baer, but armed with better microscopes and the new theory. As he compared embryos of many species, he developed a radical new hypothesis called *recapitulation.* He believed that as an individual organism undergoes its individual development (*ontogeny*), it replays the evolutionary history of its species (*phylogeny*). All life began as a single cell; so does an individual. The earliest multicellular life forms were probably ball shaped, with just a few different types of cells; a human embryo goes through a similar phase. Only in later stages of development do animal embryos start to look markedly different from one another. For Haeckel, this reflected the fact that most of today's species arose recently in evolutionary history.

Haeckel found fascinating evidence for his claims. At one stage a human embryo develops structures like gill slits, which then disappear again. This only made sense, he said, in light of the fact that the distant ancestors of mammals were fish. Haeckel drew images of embryos at various phases to show how similar their body plans were. This work has been criticized because his drawings tended to emphasize the similarities rather than the differences between embryos. Haeckel's defenders point out that such a critique is easy to make in the days of photography, where objective images can be made of samples. Drawing is subjective; the artist must make choices about which features to emphasize after looking at many specimens, and there is always a danger of wishful thinking creeping into the process.

The recapitulation hypothesis was in many ways logical and appealing. Knowing nothing of genes or DNA or their roles in shaping organisms, researchers were struggling to understand how one species might be transformed into another. It was easy to imagine that this could happen when a species added on developmental stages or its development slowed down. But Haeckel took the idea further, claiming that "ontogeny recapitulates phylogeny" was a law. A human embryo did not simply resemble that of a fish; he believed that it became a fish—an adult fish—on its way to becoming an adult human. The hypothesis claimed that evolution worked by adding new developmental stages to the end of an animal's life. This use of adult forms as a reference point was a serious flaw in his hypothesis.

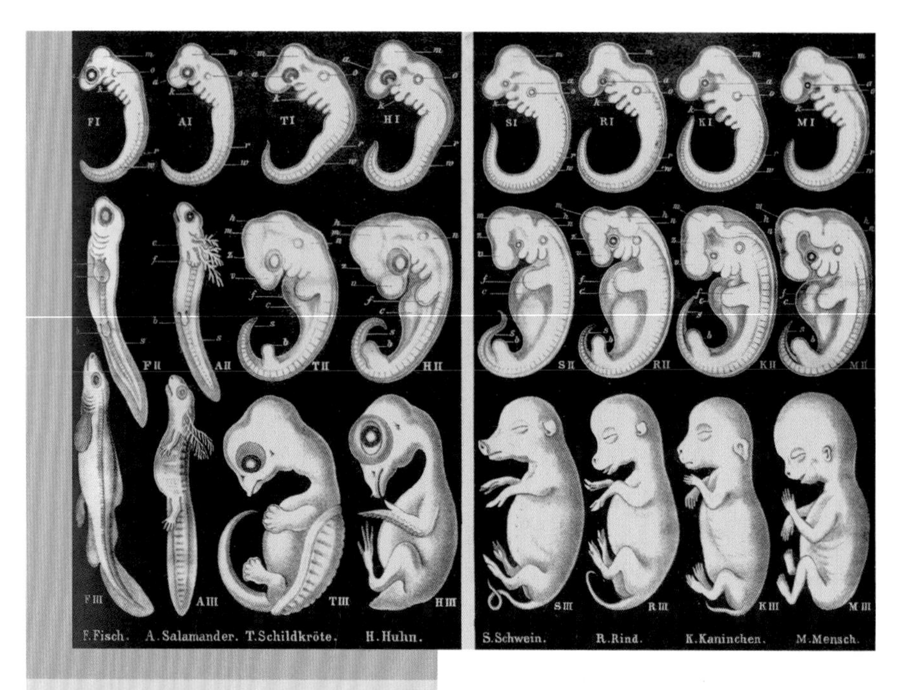

Ernst Haeckel's drawings of embryos at various stages of development emphasize the fact that many species go through phases in which they resemble one another.

This idea was appealing at the time. Darwin had firmly stated that evolution did not mean that species improved themselves; they merely changed, mostly in response to pressures from a changing environment. The public—and many scientists—were uncomfortable with the idea. Humans were obviously so much "better" than other species that the whole point of evolution must have been to produce better and better creatures until finally arriving at *Homo sapiens.* Decades before Darwin's discoveries, the Frenchman Jean-Baptiste Lamarck (1744–1829) had proposed that offspring could inherit things learned or acquired during their parents' lifetimes. Recapitulation seemed to be a way to rescue part of this idea—human embryos passed through a phase that looked like a fish on their way to becoming something better.

Haeckel's hypothesis was useful because it made embryologists everywhere start thinking in evolutionary terms. Even so, it was mostly wrong. Fritz Müller (1821–97), a German living in Brazil, studied crustaceans to prove that evolution shaped larvae as well as adults. As free-swimming organisms, the larvae of each species would have to cope with predators and other features of the environment, so natural selection would shape them just as it affected adults. As new branches developed on the tree of life, larvae would develop new features and pass them down, so there would not be a single type of larval phase preserved within every species. Numerous examples of such adaptations have been found. Species such as water fleas, frogs, and carp develop differently when predators are nearby. Water flea larvae grow larger, helmet-shaped heads that make them harder to swallow. Tadpoles grow stronger tails that allow them to swim more quickly and make faster turns.

The value of the recapitulation hypothesis at the time was that it encouraged scientists to focus on developmental processes rather than on only end results such as fully formed limbs. If the bones in a dolphin fin could be matched one to one with the hand of a primate, there also had to be a match between the processes that created the bones. This could be followed all the way back to the earliest stages of gastrulation, with the creation of three layers of tissues and structures such as the notochord and eyes. But for decades scientists were stuck there. They did not know why embryos underwent these processes, because they had not yet discovered the genetic code.

A few decades later Walter Garstang (1868–1949) and Gavin de Beer (1899–1972) pointed out the importance of timing when comparing embryos of different species. New species did not usually arise by adding on developmental stages, as Haeckel had proposed. Instead, each organ and bodily system should be looked at as an independent module. The development of one part might speed up compared to the others, a bit like the way engineers make changes in computers. They might develop a new graphics or sound card, while the rest of the machine stays the same. Of course, this may well put pressure on other parts of the machine to change; new games might be made to take advantage of the features of the graphics card, and for them to function well, they might require more RAM or changes in the keyboard. The fact that one change often prompts others could explain why a new species had longer limbs than its ancestors, or why humans and chimps do not have tails.

Recapitulation is undergoing a sort of limited revival in the molecular view of evolution. Today it might be phrased in this way: "Organisms resemble each other at many stages of development because the genes they inherited from common ancestors work in a similar way to create the same kinds of body structures." This happens even when the starting points and ending points of development are different: The eggs of a chicken and human are quite different, and they are very different as adults, but particular phases of embryonic development are similar.

Denis Duboule at the University of Geneva in Switzerland has come up with an explanation that he calls the "phylotypic hourglass." The very first stages of embryonic development are dictated by the way a species reproduces. During the final stages the embryo takes on the features that will help it survive in the outer world. The specific contexts are unique for each species, but during the middle stages an embryo usually is protected in the mother's womb. There is little pressure from the environment, so it is at this phase that organisms most resemble one another, and they also most resemble their last common ancestor. Duboule compares the situation to an hourglass, which has a wide opening at the top (the egg state); another wide opening at the bottom, showing how many different forms animals can assume; and a pinched area in the middle, representing a common phase that most pass through.

A few species have indeed evolved the way Haeckel believed, through a process that Stephen Jay Gould called "terminal addition." In this process a new species adds developmental stages beyond those of its ancestors (like adding boxcars to a train). In other cases evolution has sent species such as caterpillars off on a path that is completely different from other kinds of larvae, like trains leaving a station in different directions. And a few types of organisms underwent the opposite of what Haeckel proposed, becoming stuck at an early phase of development because of flaws in the genes that were supposed to trigger the next step.

An example is the axolotl, a rare salamander found only in Lake Xochimilco in southern Mexico City. This creature remains in its larval form its whole life and can even reproduce without becoming an adult. (Some types of axolotl can be pushed to mature further, like other salamanders, if they are given a hormone called thyroxin; their gills will recede and their lungs will expand, allowing them to leave the water.) Like many other salamanders (and the embryos of most species), the axolotl can regenerate lost or damaged parts of its body, which may have something to do with its slowed-down development.

GENOMES AND HOMOLOGS

In the late 1970s and 1980s researchers such as Nüsslein-Volhard began using genetics to study the development of embryos (described in chapter 1). Prior to that time most scientists believed that genes had undergone so many mutations they would work completely differently in each species. Very quickly, however, researchers discovered that developmentally important genes had been conserved very widely throughout the animal kingdom. Not only could they be recognized in different species, they could sometimes even be transplanted from one organism to another and still function.

And a similar defect in a gene can have the same effects on mice and humans. In 1995 Dora Games and Jun Zhao of Athena Neurosciences, a California company specialized in neurological research, worked with researchers from across the United States to develop a strain of mouse with a mutation in the APP gene, which is linked to human Alzheimer's disease. The mice were normal at birth but within a few months began to exhibit symptoms characteristic of the human disease. This was the first time the condition could be imitated in mice, and the animals have been used extensively in the search for substances that might slow down or reverse the development of symptoms.

In developing such models researchers must be sure they have found the closest animal relative of a human gene. The genes of two species are never absolutely identical because of naturally occurring mutations. It is usually easy to recognize their relationships in closely related species, such as mice, chimpanzees, and humans. This becomes harder the farther apart two species are from each other on the evolutionary tree. With very distantly related species it is often impossible to find related genes without pattern-scanning computer software.

The closest match between related genes in different species is called a *homolog.* It may not be the only match; a gene found in one species may have two or more close relatives in another. This happens because of mistakes that happen as DNA is copied, giving the offspring of an organism duplicates of genes, chromosomes, or even the complete genome. At first, an extra copy of a gene usually carries out the same tasks in cells as the original, but soon mutations create differences. They may disable one copy (turning it into a *pseudogene*) or give it new functions.

The fate of copies often depends on environmental influences and natural selection. Mice and other rodents have a large number of genes devoted to smell, many of which they acquired through gene duplications. Smell is important to a mouse in finding food and mates and avoiding predators. Mutations in the copies of these genes made the rodent nose much more sensitive and undoubtedly helped animals survive, which increased the likelihood that duplicate genes would be retained rather than lost. Humans have lost most of these genes and are much less sensitive to smell overall, but there is some evidence that they have developed genes that make them more sensitive to types of smells that are important to them, for example, of cooked food.

Until the completion of an organism's genome there is no sure way to scan to see how many copies of a gene it might have. The complete DNA sequences of humans and other organisms have turned out to be a treasure trove of information. A 2004 study by Andrew Fortna and James Sikela's group at the University of Colorado Health Sciences Center looked for gene duplications in the human genome and five other species of apes. They discovered more than 1,000 duplications that had occurred in these species. Of these, 134 were unique to humans, which meant they had occurred since humans branched off from other hominoid species. Some may have played an important role in the development of unique human characteristics. A gene called NAIP, for example, is known to prevent the death of neurons. Extra copies of NAIP might have helped preserve cells as the brain developed and increased its size. Other duplicates are known to participate in such brain functions as memory and learning.

Acquiring an extra copy of a gene usually doubles the amount of a particular protein produced by the body, which is sometimes dangerous. In 2005 Andrew Sharpe, Devin Locke, and Evan Eichler of the University of Washington School of Medicine in Seattle collected evidence showing that dozens of duplications have played a role in hereditary human diseases.

However, an extra copy of a gene can also be helpful. It may be available to step in if the original is lost or damaged. Michael Lieberman and colleagues at the Baylor College of Medicine in Houston, Texas, showed that when a gene called GGT is disrupted in mice, it causes a severe condition that leads to early death. Humans with mutations in this gene have serious symptoms, but they are usually milder than in mice, probably because they have closely related copies of the gene that are able to step in and take over the functions of the defective molecule.

In another case scientists had discovered that a molecule called TPH seemed to be involved in psychiatric disorders. Its function was to help produce serotonin, a small molecule that plays a key role in helping brain cells communicate with one another. But when TPH was deleted from mice, the animals showed no symptoms. Michael Bader's laboratory at the Max Delbrück Center for Molecular Medicine in Berlin had been studying the gene. They suspected that mice and humans might produce a second form of TPH, and now they could draw on data from both organisms' complete genome sequences to try to find it. Their search revealed a second gene in both species, TPH-2, which turned out to be the molecule that actually controls the production of serotonin. The discovery permitted the development of strains of mice that do not produce TPH-2, which are now used as models to study behavioral and psychiatric disorders in humans.

Geological maps give mining companies important information about where to search for oil and other natural resources, and maps of gene activity in various parts of the body could provide a similar function for disease researchers. Until recently gene activity had to be detected on a molecule-by-molecule basis, but now scientists have access to methods such as DNA microarrays and mass spectrometry that can scan the molecules produced in particular cells, tissues, and organs. Information from their experiments is now being combined into maps.

Pamela Hoodless and Marco Marra at the Canadian National Cancer Institute in Vancouver have created a database called the Mouse Atlas of Gene Expression in order to track the genes that are expressed in different tissues over the course of development. The project, which was launched in 2007 with a budget of $13.2 million, will allow researchers all over the world to compare the functions of mouse genes to those of other organisms and develop better models of disease. An older version, called the Edinburgh Mouse Atlas, has a very interesting interface that allows users to watch the use of genes over the course of mouse development.

HOX GENES AND GENOME EVOLUTION

Hox genes were introduced in chapter 3 as master patterning genes that evolved long ago and are responsible for creating the head-to-tail "scaffold" of animal bodies. Starting in the 1970s the laboratories of Gehring in Basel and Kaufmann in Indiana began studying their functions and using them to understand animal evolution. While the developing embryos of flies and humans look quite different, the homeobox genes that trigger the formation of bodies are so similar that Jonathan Slack of Oxford University has called them a "Rosetta stone" of development. They are excellent examples of the way evolution uses existing elements to spin off new patterns and species often markedly different from their ancestors but drawing on a common underlying plan. That developmental program has survived virtually

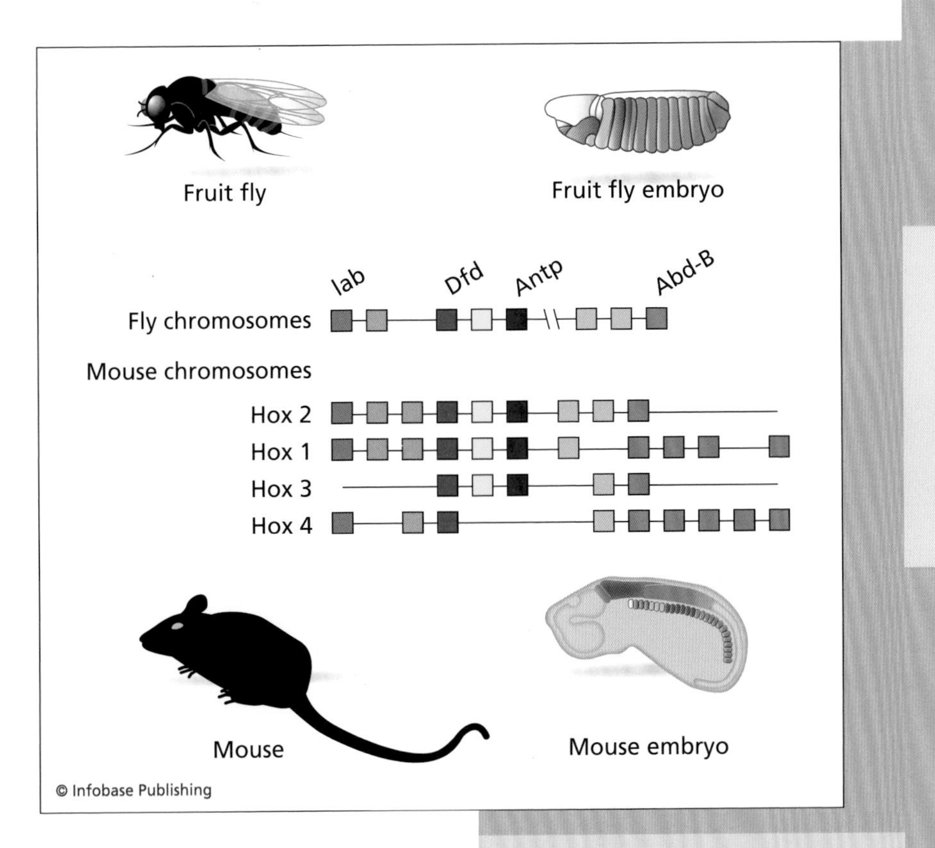

As an embryo develops, Hox genes are switched on in a precise order to control the growth and formation of body structures in the head-to-rear direction. These genes have very similar functions and are activated in the same basic order in species of animals ranging from insects to humans.

intact as species have evolved over hundreds of millions of years.

The Hox genes have become crucial pieces of evidence in resolving some questions about evolution, particularly regarding early events in the rise of vertebrates. The homeobox cluster in vertebrates typically contains 13 genes, and the cells of humans, mice, and many other vertebrates contain four copies of the cluster. This fact led Susumu Ohno (1928–2000) to propose that in some of the earliest vertebrates, the entire genome had been duplicated—probably twice. There were various scenarios by which this might have happened: A fertilized egg might have divided without separating

An Evolutionary Tool Kit

Sean Carroll, who is a professor at the University of Wisconsin and a recognized expert in the evo-devo field, has helped to prove that Hox genes were present in the common ancestor of insects and vertebrates and that they serve as master genes to produce limbs. Some of his recent work has been based on studies of onychophorans, wormlike creatures that also have characteristics of insects. Two things in particular about these animals interested him: First, they separated off from the branches that led to arthropods (insects, crustaceans, and myriapods) early in animal evolution, probably in the Cambrian period. Rocks from this geological period, which began about 542 million years ago and lasted about 55 million years, contain an amazing variety of fossils. It was during this era that evolution produced most of the types of animal bodies known today.

Any features that onychophorans shared with insects were almost sure to have existed in the common ancestor of all of these types of species. Especially interesting was the way they walked. Onychophorans have tubelike appendages called "lobopods," more like stumps than legs. From studies of fossils Carroll and many others believed that the ancestors of arthropods had limbs like these and that they eventually evolved into legs, arms, wings, claws, the fangs of centipedes, and antennae. One very popular idea was that many new types of animals—and these specialized limbs—arose through the appearance of new Hox genes at the beginning of the Cambrian period, but Carroll had his doubts. A look at the genes of onychophorans might resolve the debate.

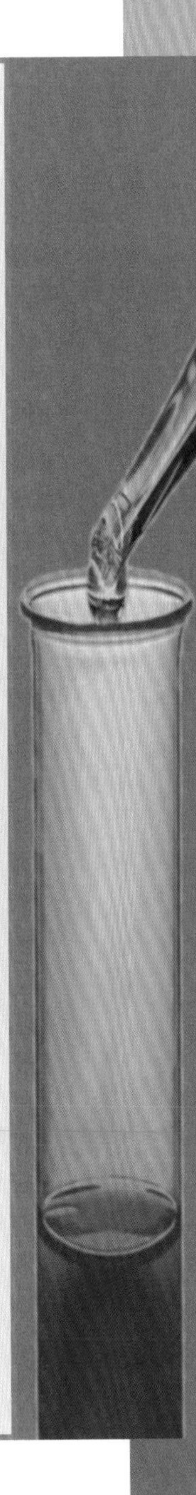

The problem was finding living examples of these animals, which existed only in a few places in the world. Carroll sent two students, Bob Warren and Jen Grenier, off to Australia, where they spent months poking through fallen logs (and avoiding poisonous spiders and centipedes) in order to collect enough specimens for an analysis. Back home in Wisconsin, the lab began a systematic investigation of the worms' Hox genes. The animals contained the same complete set found in their remote cousins, which meant that nature had not had to come up with new ones to build legs, the pincers of crabs, or the wings of birds. All of these structures belong to what Carroll calls a common genetic "tool kit." The tools were fine-tuned in different organisms, and there were shifts in the areas in which the genes were active. "We found the same mechanism in all the divisions of the animal kingdom," Carroll said. "The architecture varies tremendously, but the genetic instructions are the same and have been preserved for a very long period of time."

Sean Carroll, professor of genetics at the University of Wisconsin, has helped explain the evolutionary origins of limbs and other body features by studying unusual wormlike creatures called onchyophorans. *(Michael Forster Rothbart/University of Wisconsin–Madison)*

its DNA into two sets, or there might have been another error in cell division. A smaller but similar "sorting problem" leads to Down syndrome, in which a fertilized egg receives three copies of chromosome 21. Whole genome duplications are known to have happened in plants such as wheat and rice, possibly as a result of humans selecting plants that provide more food, and in species of fish.

Ohno proposed that the two duplications happened just as vertebrates started off on a new evolutionary branch; it might even have been the cause of their divergence. The duplications would have given the species a huge number of new genes—a vast biological toolbox for nature to improvise on. At first, the new copies probably behaved like the original, but mutations gave them slightly different functions. Many of the genes have become nonfunctional and disappeared, which obscures the evolutionary record; nevertheless, most scientists are convinced that one complete genome duplication happened. Ohno's hypothesis about the second has not yet been proven because vertebrates do not contain four copies of every gene found in other animal families. Rather than a single event there may have been several, in which large portions of DNA were copied. It should be possible to prove or disprove the hypothesis when complete genomes are available for a wider range of animals.

DNA sequencing has continued at an ever-increasing pace since the completion of the human genome; it has become easier and cheaper to finish other genomes. As information from new organisms becomes available, the Hox clusters continue to provide rich data about evolution. In turn, this is giving scientists a better idea of how the genes function.

In 2002 Cynthia Hughes from Kaufmann's lab at Indiana University investigated the activity of Hox genes in centipedes, trying to understand how the animal had evolved so many body segments. She also hoped to get a better idea of the overall placement of these animals and their relatives—called "myriapods"—on the evolutionary tree. They are closely related to the branches that produced insects and crustaceans, such as lobsters, but the exact relationships have been unclear.

The study revealed some interesting differences between how Hox genes are used in the different species. The genes in centipedes switch on in the same stepwise order in the embryo to define segments of the body from the head downward, but in flies and crustaceans one gene predominates in each segment, whereas several genes usually overlap in the centipede segments. One of these regions, called the "maxilliped," combines genes that usually appear only in the head or the next segment, the first part of the thorax, but not both. Interestingly, this mixture of codes creates a segment that also has a mix of functions. It grows extensions that first look like legs but then become poisonous fangs.

GENE HOMOLOGIES AND THE ORIGINS OF ORGANS

The birth of each organ in the body is triggered through the activation of key genes. Cerberus launches the development of a head, FGF is necessary to build an arm or leg, and Pax6 initiates the construction of eyes. Related copies of these genes are found in all animal species, where they usually stimulate the same types of processes. They are also found in single-celled organisms such as yeast. Just as Hox genes have served as molecular footprints that can be followed back to the earliest animal bodies, molecules such as Pax6 are telling researchers about the origins of the body's organs.

Critics of evolution in the 19th century claimed that extremely complex organs such as the eye could not have arisen from much simpler structures. (No one seemed to consider the fact that it happens in every embryo, where the most intricate biological "machines" arise from a single cell.) Darwin admitted that it would be difficult to retrace all the steps in the evolution of the eye, but he had no doubts that the theory could explain its structure and development. Given a single light-sensitive cell, he proposed, billions of years of natural selection could take care of the rest. Today, studies from genetics and development are providing the evidence to show that this is exactly what happened.

The eyes of flies and humans look so different that some scientists proposed they evolved independently. It was clear that insects and vertebrates shared a common ancestor, and it might have had eyes, but of which type? Did all of the eyes of today's animals arise through reshaping and refinements of this single original? Or had a second set of cells evolved to produce a completely different type of eye? There was no way to answer these questions without a detailed look at the cells making up the eyes and the molecules that governed their development.

Studies under the microscope revealed two basic types of light collectors, called "photoreceptor cells." The most obvious difference between the types was their shapes, which had evolved in fundamentally different ways. The proteins that transform light into chemical signals, opsins, are located in the membrane. So just as someone might mount more solar panels around a house to collect more power, cells have increased their sensitivity to light by increasing their surface area. Ciliary photoreceptor cells accomplished this by sprouting hairlike extensions across their surface—like filling the yard with solar panels. Rhabdomeric photoreceptors have a single large protrusion that has grown branches—like hanging the solar panels from a tree.

Rhabdomeric cells are the main light collectors in the fruit fly, whereas the rods and cones of humans and other vertebrates consist of ciliary photoreceptors. But the eyes of many species contain both types of cells. This implied that the ancestor also had both. To resolve the debate, Detlev Arendt and Jochen Wittbrodt at the European Molecular Biology Laboratory in Germany decided to look at the eyes of *Platynereis dumerilii,* a tiny marine worm. This organism is regarded as a "living fossil" because of its place on the evolutionary tree; its lineage separated from the branches that led to insects and vertebrates very soon after the development of the first true animals. Any characteristics it shares with flies or vertebrates, researchers believe, were almost surely present in their common ancestor.

Platynereis is interesting because it grows multiple sets of eyes over the course of development. One set appears in early larvae but disappears when the adult eyes come into use. Both sets of eyes capture light using rhabdomeric cells, the way fruit flies do. But as Arendt and Wittbrodt probed the surrounding tissue, they discovered another set of cells deeper inside the brain. These cells had all the characteristics of ciliary photoreceptors; they even produced opsin proteins needed in eyes. Comparing the kinds of cells, their locations, and the similarity between their genetic programs allowed the scientists to conclude that the common ancestor of *Platynereis,* humans, and insects had both types of photoreceptors.

Arendt wondered if vertebrates might still have some descendants of rhabdomeric cells, so he took a closer look at other key genes that were active in the worm's photoreceptors. When he had identified several key genes, he went looking for vertebrate cells that used them. He discovered that they are no longer used directly in the eye, but they have developed important support functions. For example, the ganglion nerves behind the retina, which pass light impulses on to the brain, are descendants of these cells. Despite their new jobs, they still are in the business of interpreting light; they help set up the body's circadian rhythm, activities such as sleeping and wakefulness that are governed by a 24-hour cycle of light and dark.

Arendt is applying the evo-devo approach to a range of similar problems. In 2007 the laboratory investigated regions of the brain that secrete tiny proteins that govern such processes as growth, development, and metabolism. Such brain centers exist in all animals, from mollusks to spiders, insects, and vertebrates. In humans the hypothalamus has this function; the comparable tissue in insects is called the "pars intercerebralis." The fact that these parts of the brain secrete similar molecules suggested that they could be traced back to the ancestor of all animals, but what type of brain did it have?

Once again the group turned to the organism *Platynereis* and used this model worm to extract a molecular "fingerprint" for the brain. They found that cells of the early embryos of worms, flies, and fish used similar genes as they developed into brain cells that could secrete small proteins. Thus light-sensitive cells in a very early wormlike creature formed a primitive brain that could receive sensory information from the marine environment and translate it into the body's responses.

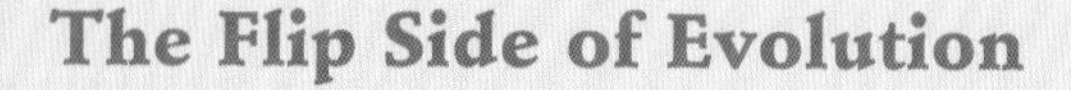

The Flip Side of Evolution

After the discovery that Hox genes organized all types of animals along the head-to-tail axis, scientists wondered whether there might be a similar master plan for creating structures along the dorsal-ventral (back-to-front) direction. One group actively pursuing an answer was that of Eddy de Robertis, next door to Walter Gehring's laboratory in Basel, Switzerland. The scientists examined genes known to help create structures in the bellies and backs of animals, but no clear parallels emerged. The major structure of the back in vertebrates—the spinal cord and nervous system—was located on the belly side of insects and other arthropods. Clearly the last common ancestor of the two branches of animals had possessed the elements needed to create a nervous system, but did it have a brain? Or eyes? No one knew how much of the developmental program for these structures it had passed along versus how much had developed independently.

Then a young student named Detlev Arendt, sitting in a library in Freiburg, Germany, had an insight that drew the pieces of the puzzle together. Arendt needed a topic for his Ph.D. dissertation. His interests included evolution and development, and he had all sorts of journals open on the desk in front of him, from very old theories about animal origins to new information about gene activation in the dorsal and ventral regions of various species. Suddenly a pattern leapt off the pages. He went to his adviser, Kerstin Nübler-Jung, who thought he might be right and encouraged him to dig deeper. He wrote up his findings and submitted them to the journal *Nature,* which took the almost unheard-of step of publishing a paper by an undergraduate student in September 1994.

To understand his discovery it is necessary to go back to the beginning of the 19th century, when the French scientist Étienne Geoffroy Saint-Hilaire proposed a radical idea. He had noticed that arthropod species (animals including insects and crustaceans) were not built all that differently from vertebrates. In fact, if one flipped insects over on their backs, the similarities became much stronger. Lacking a theory of evolution (which would not be published for nearly another four decades), Geoffroy could offer no explanation. His hypothesis languished in the anatomy books alongside other odd observations, such as the fact that the flippers of dolphins and whales were built much differently from the fins of fish.

After Charles Darwin published *On the Origin of Species*, Saint-Hilaire's ideas suddenly took on a completely new meaning. In 1875 Anton Dohrn (1840–1909), founder of a research station in Naples, suggested that the common ancestor of insects and humans might have been a wormlike animal with a heart at the back and its nerves in the stomach. It might not matter very much to a worm which side was up. Mutations could have created a body built upside down, leading to a new branch of life. The "worm" that eventually evolved into vertebrates (hundreds of millions of years later) had things backward: its nerves at the back and its heart toward the front. Dohrn's ideas seemed reasonable, and they fit well with evolutionary theory, but scientists continued to debate the question for more than a century. The hypothetical ancestor no longer existed, and insects and vertebrates had undergone so many evolutionary changes and rearrangements of their internal organs that some researchers despaired of ever figuring out the details of their evolutionary relationship.

Hox genes, however, promised to fill in some of the missing chapters. Genes might still bear traces of an

(continues)

(continued)

ancient "flip." Cases like antennapedia showed that even anatomical features that looked very different could mask a common genetic pattern, and this is what Arendt discovered that day in the library. The data produced by scientists such as Edward Lewis, Gehring, de Robertis, and many others, who had been studying the role of genes in building body structures in several organisms for more than a decade, gave Arendt a great advantage over scientists of the last century. Yet the dorsal-ventral picture was confusing, particularly since scientists were trying to combine two patterns into one without realizing it.

By assuming that there had been a flip, Arendt discovered that insects and vertebrates used the same complex of molecules—including genes called achaete and scute—to begin to build nerves in both types of organisms. As he dug through the scientific literature, he found that other genes not directly involved in the nervous system showed a similar inverted pattern of expression. Dpp helps create specialized structures in the back in flies; its closest relative in vertebrates, Bmp4, guides the formation of the ventral side. Studies of genes had suddenly breathed new life into an old and almost forgotten theory about the relationships between insects and vertebrates.

The findings had another implication, Arendt and Nübler-Jung wrote: They might change how scientists thought about the common ancestor of these different types of animals. It was likely that the genes that build the nervous system in today's animals had a similar effect on development in the distant past. Their ancient wormlike creature might well have already possessed a brain, eyes, and other sophisticated elements of the nervous system.

XENOTRANSPLANTATION AND THE CREATION OF ARTIFICIAL SPECIES

François Jacob, a pioneer of modern biology, has said, "When I started in biology in the 1950s, the idea was that the molecules from one organism were very different from the molecules from another organism. For instance, cows had cow molecules and goats had goat molecules and snakes had snake molecules, and it was because they were made of cow molecules that a cow was a cow." When scientists learned to sequence DNA, however, they discovered that this was not the case. Cow and human genomes are built of most of the same genes, partly disguised by mutations and errors that have given one species extra copies of genes. This similarity has permitted genetic engineering experiments in which genes can be swapped between species.

One reason to do this is to discover whether small differences between the versions of genes found in different species are important. Changing a single letter of the genetic code can sometimes lead to severe illness or death, but most mutations are harmless. The only way to know whether a particular mutation contributes to Alzheimer's disease, for example, may be to insert the defective human gene into a mouse. If the animal develops symptoms similar to those in humans, the gene can also be used to search for cures.

One medical application of genetic engineering is to turn cells into factories to produce substances such as human insulin or the human growth hormone, which cannot be obtained in high quantities from human donors. Researchers have even grander plans for these techniques. By inserting multiple human genes into the DNA of pigs and other animals, for example, they hope that the animals will grow organs that can be used for *xenotransplantations* in humans. The number of human organs that are available for transplants does not nearly meet the current demand—about 60 percent of patients on waiting lists die before receiving a healthy organ—so scientists have turned to animals as a possible source.

Obviously, the same problems that have been encountered with human transplantations, particularly organ rejection by the immune system, will occur even more dramatically in xenotransplantations, which is why most attempts so far have not been very successful. The earliest recorded attempts were made in 1963, when Dr. Keith Reemtsma of Tulane University in New Orleans, Louisiana, transplanted chimpanzee kidneys into 13 humans. All of the patients eventually died, but one survived for nine months, showing that animal organs might be useful as bridges to keep a patient alive until a human organ became available. A decade after Dr. Christian Barnard performed the first successful human heart transplant in 1967, he tried to use baboon and chimpanzee hearts to keep two patients alive after unsuccessful heart surgery. Both died. Soon after, however, doctors could draw on new immunosuppressive drugs such as cyclosporine, which were greatly reducing the number of human organ rejections. The substances might also work with xenotransplantations. In a famous case from 1984, surgeons transplanted a baboon heart into a human newborn named "Baby Fae," who survived for 20 days before rejecting it.

In 1992 doctors connected a patient to a pig liver kept in a plastic bag by the bedside and hooked up to her blood vessels. The woman survived until a human liver was available for transplantation. The pig's tissue would not have kept her alive much longer, however, because her body would have rejected it.

Scientists hope that the addition of human genes will cause the pigs to develop livers that can fool the human immune system. In July 1995 the U.S. Food and Drug Administration gave Duke University Medical Center in Durham, North Carolina, the green light to conduct transplantations of organs from genetically engineered pigs into a few patients with end-stage liver disease. Researchers have tried using organs from species more closely related to humans, such as baboons, but this has raised ethical and practical concerns: Primates are difficult to raise, their organs are sometimes too small to serve the human body, and diseases may lurk in their tissues that could then infect a human, particularly someone whose immune system has been depressed in order to accept a transplantation.

The field of xenotransplantation experienced a setback in 1997 when Robin Weiss at University College London discovered that the pig genome contained a *retrovirus* called PERV that might infect human cells. Retroviruses such as PERV and HIV survive in cells by slipping their own genetic code into that of the host cell. The information may lie dormant for a long time before it is activated; then the cell begins producing viruses that go on to infect other cells. The discovery of the retrovirus raised fears that PERV might jump from pig tissue to infect human cells, and a ban was placed on xenotransplantations until scientists developed a test sensitive enough to detect small traces of the virus. In the meantime, no traces of PERV have been found in patients who received pig organs, although Weiss and others proved that the virus was able to infect human cell cultures in the laboratory.

This and other concerns led many countries to pass laws regulating xenotransplantations in the late 1990s. Research has continued under strict controls and has received a great deal of media attention. Recent experiments include attempts to use pig brain cells to treat Parkinson's disease in humans, bone marrow for blood diseases, and pancreatic cells transplanted into patients with type 1 diabetes. The hope is that these cells will act like healthy human stem cells and rebuild damaged tissues. This can happen because signals from the "new neighborhoods" usually succeed in reprogramming cells to take on the functions that the tissue needs. If problems of rejection can be overcome, researchers hope that xenotransplantations of stem cells will someday become standard tools in the fight against disease.

The most dramatic case of fusing cells from different organisms involves the creation of new types of hybrid animals. This has long been done with relatively simple species, and genetic engineering experiments have transplanted increasing numbers of genes between species.

Once again, this practice raises serious ethical issues, particularly in the context of using human genes in animals. In 1997 Stuart Newman and colleagues at the New York Medical College applied for a patent to obtain ownership of some creatures that were part animal, part human. These organisms do not yet exist, but one day researchers might create something like them.

Jeremy Rifkin *(Jeremy Rifkin)*

Newman and his partner, biotechnology activist Jeremy Rifkin, were worried about experiments involving increasingly complex mixtures of the genes of humans and other species. At the University of Nevada human stem cells had been implanted into sheep embryo with the aim of obtaining livers and other organs that can be transplanted into humans. Pigs had been engineered to produce human hemoglobin, and Stanford University scientists had injected immature human brain cells into mouse fetuses, hoping to create mice that could serve as models of Parkinson's disease and other brain disorders.

But these experiments were overshadowed by an even more dramatic case. In 1984 Carole Fehilly's laboratory in Cambridge (Great Britain) and Sabine Meinecke-Tillmann and her colleagues in Germany succeeded in making "geeps"—animals that were half goat, half sheep. They did this by removing an embryo from each type of mother at an early stage, when each consisted of about eight cells, and allowing them to fuse in the test tube. The embryos were then reimplanted into a sheep, where they grew. Some of the animals produced in this way turned out to be patchworklike mixtures, with some parts of the body developing from the sheep cells and others from the goat.

In their patent application Newman and his colleagues stated that there was nothing to prevent scientists from carrying out similar projects on humans and their closest living relatives, chimpanzees or other primates. They hoped the application would force lawmakers to clear up regulations concerning genetic engineering. A patent would block other scientists from making such creatures and buy time for scientists and governments to think about the problem. The U.S. Supreme Court had set a dangerous precedent in 1980, they felt, with a decision that allowed the patenting of a microorganism; the U.S. Patent Office had moved farther down the "slippery slope" by granting a patent for a mouse strain to Harvard University.

In that case Philip Leder and his research team had genetically modified the strain by inserting a gene that caused the animals to develop cancer. This "oncomouse" has been useful to researchers all across the globe in uncovering some of the causes of cancer, developing drugs to combat it, and detecting carcinogens in the environment. When the patent was awarded, animal rights and environmental groups protested on the grounds that the animals suffered and that no one should have the right to own a species. Defenders of patenting said that without the promise of profits from ownership, companies would have little interest in doing medical research. After a pause of nearly five years, during which experts tried to untangle the legal and ethical issues surrounding the case, the Patent Office once again began issuing patents on genetically modified animals.

The application made by Rifkin and Newman was rejected on the grounds that it proposed creating animals that were "too human." They had succeeded, meanwhile, in calling attention to such problems as where to draw the line between human and nonhuman and why some forms of life should be ownable but others not. These questions are not well defined in current laws or patent practices. But they will need to be in order to deal with the increasing number of ethical issues that surround genetic engineering.

SUMMARY AND SOME FINAL THOUGHTS

Evolutionary developmental biology has brought together several fields of science that for a long time were practiced with different tools and were limited to asking certain types of questions. Charles Darwin's theory explained how new species arose and how they were related to one another and even provided some ideas about the origins of life itself. In the middle of the 20th century the discovery of DNA's structure allowed scientists to begin to read the genetic code—and to track the changes that had occurred over the course of evolution. By the 1980s and 1990s researchers had created powerful tools to manipulate genes and watch their influence on the form and behavior of organisms. Modern developmental biology, which arose in the 1980s, began to pull all of these pieces together by showing how the activity of genes guided the differentiation of cells and the formation of tissues. Evo-devo draws these elements together by exploring how related genes create similar forms in different species and how slight changes in genes create the differences between species.

Developmental biology will only be "complete" when scientists understand all the molecular triggers that transform a fertilized egg into all of the body's types of cells, when they have decoded the signals that guide tissues as they migrate through the body, and when they understand how the body develops as a dialogue between the genes in cells and the environment. The last few decades have seen great progress in addressing each of these questions, but many have yet to be answered. Scientists still do not know enough about the relationships between complex sets of genes and development. Ideally a researcher would be able to look at the sequence of a genome and predict what sort of organism it would produce. Scientists are still far from that goal. Today it is fairly straightforward to study what happens when one or two genes are defective in a model organism such as the mouse, but most of what happens in an animal is the result of an interplay between hundreds or thousands of genes. Coping with that complexity will be the great challenge for the next generation of biologists.

Chronology

1651	William Harvey claims that all animals arise from eggs.
1669	Niels Stensen, working for the Medici family in Florence, proposes that strata in rock reflect geological time, with the oldest layers at the bottom and newer layers above. He recognizes that fossils are remains of living creatures.
1677	Anton van Leeuwenhook discovers sperm.
1694	Rudolph Camerarius identifies the sex organs of plants.
1740	Carl Linnaeus begins to pollinate plants artificially.
1751	Pierre-Louis Maupertuis studies polydactyly, the inheritance of extra fingers in humans.
	Joseph Adams recognizes the negative hereditary effects of inbreeding.
1802	Jean-Baptiste Lamarck publishes *Research on the Organization of Living Bodies,* in which he claims that species change through environmental influences and through their own activities and behavior. His philosophy of species change is based on the notion that animals seek to

	become more perfect and better adapted to their surroundings.
	William Paley publishes the monumental work *Natural Theology,* which examines natural phenomena in order to reveal God's design of the world.
1817	Georges Cuvier publishes *The Animal Kingdom,* in which he argues that all the characteristics of a species are attuned to fit its lifestyle. His studies of fossils show that they were living creatures that have become extinct.
1820s	Étienne Geoffroy Saint-Hilaire's studies of animal anatomy point out unusual relationships between species and features that seem to have lost their functions over time.
1824	Joseph Lister builds a new type of microscope that removes distortion and greatly increases resolution.
1827	Karl Ernst von Baer is first to discover an egg cell in a mammal (a dog).
1830	Giovanni Amici discovers egg cells in plants.
1838	Matthias Schleiden states that plants are made of cells.
1840	Theodor Schwann states that all animal tissues are made of cells.
1856	Gregor Mendel begins experiments on heredity in pea plants.

1857 Joseph von Gerlach discovers a new way of staining cells that reveals their internal structures.

1858 Papers by Charles Darwin and Alfred Russel Wallace regarding evolution by natural selection are read at a meeting of the Linnean Society in London. Neither man is present.

Rudolf Virchow states the principle of *Omnis cellula e cellula*: Every cell derives from another cell—including cancer cells.

1859 Charles Darwin publishes *On the Origin of Species.* The first printing sells out on the first day.

1865 Gregor Mendel presents his paper "Experiments in Plant Hybridization" in meetings of the Society for the Study of Natural Sciences in Brnø, Moravia. The paper outlines the basic principles of the modern science of genetics.

1868 Fredrich Miescher isolates DNA from the nuclei of cells; he calls it "nuclein."

1876 Oscar Hertwig observes the fusion of sperm and egg nuclei during fertilization.

1879 Walther Flemming observes the behavior of chromosomes during cell division.

1885 August Weismann states that organisms separate reproductive cells from the rest of their bodies, which helps explain why Jean-Baptiste

Larmack's concept of evolution and inheritance is wrong. He tries and fails to observe Lamarckian inheritance in the laboratory by cutting off the tails of mice for many generations.

1894 Emil Fischer describes the fact that specific enzymes recognize and change each other using the metaphor of locks and keys.

1900 Hugo de Vries, Carl Correns, and Erich von Tschermak-Seysenegg independently publish papers that confirm Gregor Mendel's principles of heredity in a wide range of plants.

Archibald Garrod identifies the first disease that is inherited according to Mendelian laws, which means that it is caused by a defective gene.

Theodor Boveri demonstrates that different chromosomes are responsible for different hereditary characteristics.

1902 William Bateson popularizes Gregor Mendel's work in a book called *Mendel's Principles of Heredity: A Defense.*

Archibald Garrod discovers that a form of arthritis follows a Mendelian pattern of inheritance, which means that it is a genetic disease.

1903 Walter Sutton connects chromosome pairs to hereditary behavior, demonstrating that genes are located on chromosomes.

1905	Nettie Stevens and Edmund Wilson independently discover the role of the X and Y chromosomes in determining the sex of animal species.
1906	William Bateson discovers that some characteristics of plants depend on the activity of two genes.
1908	Archibald Garrod shows that humans with an inherited disease are lacking an enzyme (a protein), demonstrating that there is a connection between genes and proteins.
1909	William Bateson coins the term *genetics.*
	Thomas Hunt Morgan discovers the first mutations in fruit flies, *Drosophila melanogaster,* bred in the laboratory. This leads to the discovery of hundreds of new genes over the next decades.
1911	Thomas Hunt Morgan discovers some traits that are passed along in a sex-dependent manner and proposes that this happens because the genes are located on sex chromosomes. He proposes the general hypothesis that traits that are likely to be inherited together are located on the same chromosome.
1913	Alfred Sturtevant constructs the first genetic linkage map, allowing researchers to pinpoint the physical locations of genes on chromosomes.

1920 Hans Spemann and Hilde Proescholdt Mangold begin a series of experiments in which they transplant embryonic tissue from one species to another. The scientists show that particular groups of cells they call "organizers" send instructions to neighboring cells that change their developmental fates.

1927 Hermann Muller shows that radiation causes mutations in genes that can be passed down through heredity.

1928 Fredrick Griffith discovers that genetic information can be transferred from one bacterium to another, hinting that hereditary information is contained in DNA.

1931 Archibald Garrold proposes that diseases can be caused by a person's unique chemistry; in other words, genetic diseases may be linked to defects in enzymes.

Barbara McClintock shows that as chromosome pairs line up beside each other during the copying of DNA, fragments can break off one chromosome and be inserted into the other in a process called recombination.

1934 Calvin Bridges shows that chromosome bands can be used to pinpoint the exact locations of genes.

1935 Nikolai Timofeeff-Ressovsky, K. Zimmer, and Max Delbrück publish a groundbreaking work on the structure of genes that proposes that mutations alter the chemistry and structure of molecules.

Calvin Bridges and Hermann Muller discover independently that a fly mutation called "bar" is caused by the duplication of a gene.

1937 George Beadle and Boris Ephrussi show that genes work together in a specific order to produce some features of fruit flies.

1940 George Beadle and Edward Tatum prove that a mutation in a mold destroys an enzyme and that this characteristic is inherited in a Mendelian way, leading to their hypothesis that one gene is related to one enzyme (protein), formally proposed in 1946.

1944 Oswald Avery, Colin MacLeod, and Maclyn McCarty show that genes are made of DNA.

Erwin Schrödinger publishes *What Is Life?*

1950 Barbara McClintock publishes evidence that genes can move to different positions as chromosomes are copied.

Erwin Chargaff discovers that in DNA samples from different organisms, the base adenine (A) always occurs in the same amounts as thymine (T) and that the same is true for guanine (G) and cystine (C).

1953 James Watson and Francis Crick publish the double-helix model of DNA, which explains both how the molecule can be copied and how mutations might arise. In the same issue of the journal *Nature,* Rosalind Franklin and Maurice Wilkins publish X-ray studies that support the Watson-Crick model. This launches the field of molecular biology that shows, over the next 20 years, how the information in genes is used to build organisms.

1958 Francis Crick describes the "central dogma" of molecular biology: DNA creates RNA creates proteins. He challenges the scientific community to figure out the molecules and mechanisms by which this happens.

1961 Sydney Brenner, François Jacob, and Matthew Meselson discover messenger RNA as the template molecule that carries information from genes into protein form.

Francis Crick and Brenner suggest that proteins are made by reading three-letter codons in RNA sequences, which represent three-letter codes in DNA.

M. W. Nirenberg and J. H. Matthaei use artificial RNAs to create proteins with specific spellings, helping them learn the complete codon spellings of amino acids.

1966 Marshall Nirenberg and H. Gobind Khorana work out the complete genetic code, the DNA recipe for every amino acid.

1968 The U.S. Supreme Court declares laws against the teaching of evolution in schools unconstitutional.

1970 Hamilton Smith and Kent Wilcox isolate the first restriction enzyme, a molecule that cuts DNA at a specific sequence, which will become an essential tool in genetic engineering.

1972 Paul Berg creates a new gene in bacteria using genetic engineering.

Janet Mertz and Ron Davis use restriction enzymes and DNA-mending molecules called ligases to carry out the first recombination: the creation of an artificial DNA molecule.

1973 Stanley Cohen, Annie Chang, Robert Helling, and Herbert Boyer create the first transgenic organism by putting an artificial chromosome into bacteria.

1975 Edward Southern creates Southern blotting, a method to detect a specific DNA sequence in a person's DNA; the method will become crucial in genetic testing and biology in general.

Cesar Milsein, Georges Kohler, and Niels Kai Jerne develop a method to make monoclonal antibodies.

1977 Walter Gilbert and Allan Maxam develop a method to determine the sequence of a DNA molecule. Fredrick Sanger and colleagues

	independently develop another very rapid method for doing so.
	Phil Sharp, Louise Chow, Rich Roberts, and Pierre Chambon discover that genes can be spliced, leaving out information in the DNA sequence as a messenger RNA molecule is made.
	Walter Gilbert and Frederick Sanger devise new methods to analyze the sequence of DNA, launching the age of high-throughput DNA sequencing.
	Phillip Sharp and colleagues discover introns, information in the middle of genes that does not contain codes for proteins and must be removed before an RNA can be used to create a protein.
	Sanger finishes the first genome, the complete nucleotide sequence of a bacteriophage.
1978	Recombinant DNA technology is used to create the first human hormone.
1980	Christiane Nüsslein-Volhard and Eric Wieschaus discover the first patterning genes that influence the development of the fruit fly embryo, bringing together the fields of developmental biology and genetics.
1981	Three laboratories independently discover oncogenes, proteins that lead to cancer if they undergo mutations.

1983 Walter Gehring's laboratory in Basel, Switzerland, and Matthew Scott and Amy Weiner, working at the University of Indiana, independently discover Hox genes, master patterning molecules for the creation of the head-to-tail axis in animals as diverse as flies and humans.

1985 Kary Mullis develops the polymerase chain reaction, a method that rapidly and easily copies DNA molecules.

1987 The U.S. Department of Energy proposes a project to sequence the entire human genome.

1988 The Human Genome Project is launched by the U.S. Department of Energy and the National Institutes of Health, with the aim of determining the complete sequence of human DNA.

1989 The Human Genome Organization (HUGO) is founded.

Alec Jeffreys discovers regions of DNA that undergo high numbers of mutations. He develops a method of "DNA fingerprinting" that can match DNA samples to the person they came from and can also be used in establishing paternity and other types of family relationships.

1990 W. French Anderson carries out the first human gene replacement therapy to treat an immune system disease in four-year-old Ashanti DeSilva.

1995 Walter Gehring's laboratory shows that a gene called Pax-6 triggers the development of eyes in insects and vertebrates, evidence that very different types of eyes had a common evolutionary origin.

The Institute for Genomic Research announces the completion of the entire genome sequence of the first complete cell: the bacterium *H. influenzae.*

The U.S. Food and Drug Administration approves the first protease inhibitor for use in treating AIDS.

1997 Ian Wilmut's laboratory at the Roslin Institute produces Dolly, a sheep, the first cloned mammal.

1998 The genome of the first multicellular organism, the worm *C. elegans,* is completed.

2000 Researchers from HUGO and Celera Genomics announce the completion of a "working draft" of the entire human DNA sequence. The complete genome is published in 2003.

The genome of the fruit fly, *Drosophila melanogaster,* is completed.

2001 Celera Genomics announces the first complete assembly of the human genome.

2002	The mouse genome is completed.
2004	Scientists in Seoul, South Korea, announce the first successful cloning of a human being, a claim which is quickly proven to be false.
2006	Shinya Yamanaka's group in Kyoto, Japan, creates the first induced pluripotent stem cells from fibroblasts.
2007	Shinya Yamanaka and James Thomson of the University of Wisconsin–Madison develop induced pluripotent stem cells from human fibroblasts.
2008	Samuel Wood of the California company Stemagen successfully uses his own skin cells to produce clones, which survive five days.
2009	Newly elected president Barack Obama lifts the federal ban on funding certain types of embryonic stem cell research.

Glossary

adenosine triphosphate (ATP) a compound containing three phosphate groups that is broken down by enzymes to provide energy for chemical reactions in the cell

angiogenesis the process by which new blood vessels form

animal hemisphere the half of an egg or early embryo that contains the least yolk, usually causing cells in that hemisphere to divide more quickly than in the other half, called the vegetal hemisphere

antibody a molecule produced by white blood cells in a random process, after a complex rearrangement of genes, by docking onto antigens, antibodies provoke an immune response

antigen a molecule that is recognized by an antibody

apoptosis a cellular self-destruct program that helps sculpt tissues as an embryo develops and destroys some types of diseased cells

archenteron a cavity formed during gastrulation through an inward migration of cells; it will form the primitive gut and then the digestive tract.

blastocoel the cavity of fluid inside the blastula

blastocyst the stage of development after the blastula; an inner mass of cells collects in the blastocoel; they will become the embryo and the cells on the outside of the sphere will become the placenta

blastomere a cell formed by cleavage in the early embryo

blastula the hollow ball of cells produced through cleavage of the first embryonic cells

cell cycle the sequence of events by which a cell divides

centrosome a small cellular structure, also known as a "microtubule organizing center," that acts as a pole of the mitotic spindle during most types of cell division

chemotaxis a process by which cells migrate in response to a chemical signal. Often the cell senses changes in the concentration of a molecule to find its way.

chimera a single complex organism whose cells have arisen from two complete, separate genomes

choanocyte cells in sponges that are closely related to choanoflagellates and use their flagella to provide the sponge with food from the water

choanoflagellate a unicellular organism with an oval-shaped cell body and a flagellum used to trap bacteria or other sources of food

chromosome large, compressed clusters of DNA and many other molecules found in the cell nucleus

cleavage the earliest rounds of cell division in the zygote, which produce the blastula

clone a molecule, cell, or organism whose DNA sequence is identical to that of another individual member of its species

cytokinesis the process by which non-nuclear parts of the cell—the cytoplasm and plasma membrane—split into two daughters during cell division

cytoplasm the largest compartment of the cell; it surrounds the nucleus and contains the cell's organelles.

cytoskeleton a system of protein tubes and fibers that gives the cell shape and structure and plays a key role in processes such as cell division and migration

diploid a cell that has pairs of chromosomes

DNA (deoxyribonucleic acid) the molecule that contains the hereditary information of all species. DNA encodes RNA molecules, many of which are used to produce proteins.

DNA chip a set of probes made of DNA, usually mounted on a glass slide, that detects RNA molecules made by different types of cells and thus allows a comparison of their active and silent genes

dopamine a neurotransmitter found in the substantia nigra and other regions of the brain that influences cognition, learning, sleep, moods, and other mental states

ectoderm the outermost of the three main layers of tissue that arise during the process of gastrulation in early embryonic development; also, the tissues that arise from this layer

embryo the stage of growth (in humans) between the zygote and fetal stage, starting about the eighth week of life, as primitive organs begin to form

endoderm the innermost of the the three main layers of tissue that arise during the process of gastrulation in early embryonic development; also, the tissues that arise from this layer

enzyme a protein that carries out certain types of chemical reactions, such as cutting other molecules or activating them

epiblast tissue in the very early embryo that differentiates into ectoderm, mesoderm, and endoderm during gastrulation

fallopian tube one of two very thin canals in female mammals that connect the ovaries to the uterus

follicle a spherical structure in the female reproductive organs containing an egg and other cells that assist in its nurture and maturation

fraternal twin one of a pair of humans who are born at the same time and share the same parents but have different genomes because they are the product of the fusion of different sperm with different eggs

gastrulation a stage in very early embryonic development in which cell divisions and migrations lead to the formation of three layers of tissue: ectoderm, mesoderm, and endoderm

gene a sequence of nucleotides in a DNA molecule that holds the information needed by a cell to create a protein

genome the complete set of DNA in an organism. The term is often used to refer to a "representative" set of genes from a species, as in "the human genome," although each individual's genome is slightly different.

hematopoeitic cell a type of stem cell that develops into one of the many types of blood cells

hermaphrodite an organism that has the reproductive organs of both a male and female

homeobox a sequence found in many genes that encodes a DNA-binding protein

homeotic gene (Hox gene) a gene that contains a code called a "homeobox," which gives the gene the ability to activate other genes. Hox genes evolved in an early animal and play a crucial role in building the head-to-tail body structure of nearly all animal embryos. They also contribute to the development of structures in limbs and many other tissues.

homolog one of a pair of related genes, cells, organs, or bodies that has been inherited from a common ancestor

hypoblast a group of cells that forms from the inner cell mass of the blastocyst and will go on to contribute to the endoderm tissue

identical twin one of a pair of humans who have identical genomes because they arose from a single fertilized egg

induced pluripotent stem cell (iPSC) a cell capable of differentiating into many types of daughters, artificially made from a specialized cell taken from a developed animal

induction the process by which molecules from one cell or tissue trigger the activation of a gene in another

in vitro fertilization the fertilization of an egg by a sperm, carried out in a test tube by a scientist

keratinocyte the main type of cell in the epidermis, produced in the lower layers, which moves upward to the surface of the skin as it develops

ligand a small molecule that binds to a receptor or another protein

major histocompatibility complex (MHC) proteins that are combined with fragments of viruses or bacteria or other foreign substances and moved to the surface of cells so that they can be discovered by immune system cells

meiosis a special type of cell division leading to the creation of germ cells, leaving them with one chromosome of each pair rather than two

mesoderm the middle of the the three main layers of tissue that arise during the process of gastrulation in early embryonic development; also, tissues that arise from this layer

metastasis a process by which a cancer cell leaves a tumor, invades a tissue somewhere else in the body, and begins to form a new tumor

microcephaly a developmental disorder caused by a mutation in the microcephalin gene or sometimes by maternal alcoholism, disease, or other problems during pregnancy; it is marked by the development of an unusually small head and brain.

microRNA a small RNA molecule produced by cells whose function is to block the use of another RNA to make proteins

microtubule a fiber built of tubulin proteins that plays a key role in the transport of molecules through the cell and is used to build the mitotic spindle during cell division. Microtubules are a major part of the cytoskeleton, the scaffold of proteins that gives a cell its shape and structure.

mitochondrion an organelle found in the cytoplasm of animal cells that acts as a "power plant," providing the cell with energy. Mitochondria are thought to have evolved from a bacterium that invaded another type of cell.

mitotic spindle a structure built during cell division; it is made of microtubules, and its function is to separate chromosomes into two equal sets

mutation a change that happens in DNA when it is not perfectly copied. This can involve the swap of one base pair for another, the insertion of extra nucleotides, or larger rearrangements of material in chromosomes.

myogenesis the process of skeletal muscle development

natural selection the process by which the characteristics of some members of a species make them more successful than others at passing their DNA on to the next generation. If the effect is very strong (for example, if one type of organism can survive and reproduce and others cannot at all) or if it lasts for many generations, the result is likely to be the development of a new species.

neural crest a short-lived strip of cells that lies between the dorsal ectoderm and the neural tube and gives rise to the autonomic nervous system, some types of muscle, and other body structures

neurotransmitter a small molecule released by a nerve cell that passes information from a neuron to another cell across synapses

neurulation an early stage in the development of the central nervous system, triggered by signals passing between the notochord and ectoderm, resulting in the formation of the neural plate and neural tube

notochord a rodlike structure that arises early in the development of a wide variety of animals. It creates a head-to-tail axis during early development, and in humans it eventually develops into part of the spinal cord.

ontogeny an individual's biological development from a fertilized egg to adulthood

oocyte an unfertilized egg

opsin a type of receptor protein that transforms photons into electrochemical signals, forming the basis of eyesight in animals

ovulation the process by which a female follicle ruptures and releases an egg, which can then be fertilized

ovum a mature, unfertilized egg of an animal or plant

parthenogenesis a type of reproduction in which an egg develops into an animal without being fertilized by a sperm. Only the mother's genetic material is used to create offspring. Parthenogenesis is a natural form of cloning found in plants such as blackberries and a few types of lizards, insects, and other animals.

phosphorylation a chemical modification of a molecule involving the addition of phosphate groups. This process is frequently used to pass signals within cells.

phylogeny the study of the relationships between ancestral species and their descendants; also a tree diagram that shows these relationships

precursor cell a kind of cell that is not fully specialized but is further developed than a stem cell, usually producing one type of mature cell (or just a small number of types)

primitive streak a furrow that forms in early embryonic development through an inward migration of epiblast cells, which will go on to form the endoderm and mesoderm

protein a molecule made up of subunits called amino acids, synthesized by cells using information in genes. Proteins are often called the "worker molecules" of the cell because of the many different functions they perform.

pseudogene a DNA sequence that in the past contained a gene (a protein-encoding sequence) but has since undergone mutations and is no longer used for the synthesis of proteins

recapitulation an outdated hypothesis that states that the development of an individual organism "replays" the evolutionary history of its species

recombination a process by which a fragment of DNA breaks away from a chromosome and then is joined to another DNA molecule

retrovirus a type of virus whose genetic material is composed of DNA; it reproduces by "reverse transcribing" the RNA into DNA, which is inserted into the genome of the host and copied when the host cell reproduces

RNA (ribonucleic acid) a molecule made of nucleotide subunits that is made through the transcription of information contained in DNA. There are different types, including messenger RNA molecules, which are transported out of the cell nucleus and are used as patterns to build proteins. Other types are involved in building proteins or in controlling whether other RNAs are used to do so.

satellite cell a type of stem cell that lies on the border of skeletal muscle fibers and is used to repair damaged muscle tissue

skeletal muscle striated muscle tissue made of fused muscle cells arranged in thick fibers, permitting voluntary movement

small interfering RNA (siRNA) artificial RNA molecules designed to dock onto specific messenger RNAs and block (or partially block) their translation into proteins

somite a short-lived structure made of mesodermal cells that arises next to the notochord and is the source of cells that give rise to some of the body's bones, skeletal muscles, and other tissues

stem cell a generic cell in multicellular organisms that can develop into a variety of subtypes

T cell a kind of white blood cell that develops in the thymus and bears receptors capable of recognizing foreign molecules and triggering an immune response by the body; there are several types, which act at different stages of an infection

telomerase an enzyme that adds DNA sequences to telomeres, at the ends of chromosomes, slowing down the process by which chromosomes become shortened each time the DNA is copied

telomere a region at the end of chromosomes containing no genes but that plays an important role in protecting the chromosome from damage as DNA is copied

teratogen a substance that triggers abnormal development of an embryo's organs, limbs, or tissues

transcription factor a protein that binds to DNA and activates or deactivates a gene

transposition a process by which genes jump from their original position in an organism's genome to another location, sometimes on a different chromosome

vegetal hemisphere the half of an egg or early embryo that contains the most yolk, usually causing cells in that hemisphere to divide more slowly than in the other half, called the animal hemisphere

vitelline envelope a protective membrane surrounding the eggs of insects and invertebrates whose function is to ensure that only one sperm of the right species fertilizes the egg

X chromosome a sex chromosome; in humans, fruit flies, and certain other organisms, females have two X chromosomes and males only have one

xenotransplantation the transplantation of cells or tissues between two organisms of different species

Y chromosome the male sex chromosome in humans, fruit flies, and certain other organisms

yolk proteins that feed an egg during the earliest stages of embryonic development. In some species, such as birds, the yolk is concentrated in one hemisphere of the egg; in others, such as humans, it is equally distributed.

zona pellucida a thick envelope of proteins and other molecules surrounding the eggs of mammals that allows the sperm to bind and to penetrate the egg

zygote a fertilized egg

Further Resources

Books

Ball, Philip. *Stories of the Invisible: A Guided Tour of Molecules.* Oxford: Oxford University Press, 2001. A very accessible book that introduces nonscientists to basic concepts in the study of inorganic and organic molecules.

Beadle, George W. "Genetics and Metabolism in Neurospora." *Physiological Reviews* 25 (1945): 660. A classic paper in which Beadle summarizes studies of the slime mold, showing that genes have a one-to-one relationship to proteins.

Branden, Carl, and John Tooze. *Introduction to Protein Structure.* 2d ed. New York: Garland Publishing, 1999. A university-level book introducing students to the basic elements of protein structure and protein families.

Brown, Andrew. *In the Beginning Was the Worm.* London: Pocket Books, 2004. The story of an unlikely model organism in biology—the worm *C. elegans*—and the scientists who have used it to understand some of the most fascinating issues in developmental biology.

Browne, Janet. *Charles Darwin: The Power of Place.* New York: Knopf, 2002. The second volume of the "definitive" biography of Charles Darwin.

———. *Charles Darwin: Voyaging.* Princeton, N.J.: Princeton University Press, 1995. The first volume of the "definitive" biography of Charles Darwin.

Caporale, Lynn Helena. *Darwin in the Genome: Molecular Strategies in Biological Evolution.* New York: McGraw-Hill, 2003. A new look at variation and natural selection based on discoveries from the genomes of humans and other species, written by a noted biochemist.

Carlson, Elof Axel. *Mendel's Legacy: The Origin of Classical Genetics.* Cold Spring Harbor, N.Y.: Cold Spring Harbor Laboratory Press, 2004. An excellent, easy-to-read history of genetics, from Gregor Mendel's work to the 1950s. Carlson explains the relationship between cell biology and genetics especially well.

Caudron, Maïwen, et al. "Spatial Coordination of Spindle Assembly by Chromosome-Mediated Signaling Gradients." *Science* 5,739 (2005): 1,373–1,376. A study of the factors that regulate the self-assembly and organization of the mitotic spindle.

Chambers, Donald A. *DNA, the Double Helix: Perspective and Prospective at Forty Years.* New York: New York Academy of Sciences, 1995. A collection of historical papers from major figures involved in the discovery of DNA, with reminiscences from some of the authors.

Crick, Francis. *What Mad Pursuit: A Personal View of Scientific Discovery.* New York: Basic Books, 1988. Crick's account of several decades of scientific work, from the discovery of the structure of DNA through years of working out the details of the "central dogma" of molecular biology.

Darwin, Charles. *The Descent of Man.* Amherst, N.Y.: Prometheus, 1998. In this book, originally published 12 years after *On the Origin of Species,* Darwin outlines his ideas on the place of human beings in evolutionary theory.

———. *On the Origin of Species.* Edison, N.J.: Castle Books, 2004. The book in which Darwin first thoroughly outlined the theory of evolution and collected thousands of facts in support of it.

———. *The Voyage of the* Beagle. London: Penguin Books, 1989. Darwin's account of his five-year voyage around the world, in which he collected specimens of many new species and facts on geology, biology, and the environment that would be crucial in formulating the theory of evolution.

Elliott, William H., and Daphne C. Elliott. *Biochemistry and Molecular Biology.* New York: Oxford University Press, 1997. A university-level overview of how the information in genes is used in cells, particularly strong in its descriptions of metabolism and signaling pathways.

Fruton, Joseph. *Proteins, Enzymes, Genes: The Interplay of Chemistry and Biology.* New Haven, Conn.: Yale University Press, 1999. An extremely detailed history of biochemistry from the late 18th century to the latter half of the 20th century, best suited for students of chemistry with an interest in learning about the history and personalities behind the main discoveries in their field.

Gavin, Anne-Claude, et al. "Functional Organization of the Yeast Proteome by Systematic Analysis of Protein Complexes." *Nature* 415 (2002): 141–147. The first comprehensive study of the protein machines at work in yeast cells, revealing hundreds of new protein complexes and new aspects of how the cell assembles and uses them.

Gilbert, Scott. *Developmental Biology.* 6th ed. Sunderland, Mass.: Sinauer Associates, 2000. An excellent college-level text on virtually all aspects of developmental biology, accompanied by a multimedia CD-ROM and supplemented by a Web site with a wealth of additional materials (see "Web sites" below).

Goldsmith, Timothy H., and William F. Zimmermann. *Biology, Evolution, and Human Nature.* New York: Wiley, 2001. Life from the level of genes to human biology and behavior.

Goodsell, David S. *The Machinery of Life.* New York: Springer-Verlag, 1993. David Goodsell is an associate professor of molecular biology at the Scripps Institute and an extremely talented illustrator. This book provides a new view of the molecular world, showing the cell as a dynamic place full of interacting, intricately folded molecules.

Hall, Michael N., and Patrick Linder, eds. *The Early Days of Yeast Genetics.* Cold Spring Harbor N.Y.: Cold Spring Harbor Laboratory Press, 1993. A collection of essays by early pioneers in genetics, with personal anecdotes and firsthand experiences from the leading figures in the study of one of the most important modern laboratory organisms.

Henig, Robin Marantz. *A Monk and Two Peas.* London: Weidenfeld & Nicolson, 2000. A popular, easy-to-read account of Gregor Mendel's work and its impact on later science.

Jablonka, Eva, and Marion J. Lamb. *Evolution in Four Dimensions: Genetic, Epigenetic, Behavioral, and Symbolic Variation in the History of Life.* A creative, entertaining, and controversial view of the connections between genetics, evolution, development, and human behavior.

Judson, Horace Freeland. *The Eighth Day of Creation: Makers of the Revolution in Biology.* New York: Simon & Schuster, 1979. A comprehensive history of the science and people behind the creation of molecular biology, from the early 20th century to the 1970s.

Keller, Evelyn Fox. *A Feeling for the Organism: The Life and Work of Barbara McClintock.* San Francisco: W. H. Freeman, 1983. A fascinating up-close account of the life and discoveries of Barbara McClintock, written just before she won a Nobel Prize for her work on "jumping genes" and other aspects of genetics.

Kohler, Robert E. *Lords of the Fly:* Drosophila *Genetics and the Experimental Life.* Chicago: University of Chicago Press, 1994. The story of Thomas Hunt Morgan and his "disciples," whose discoveries regarding fruit fly genes dominated genetics in the first half of the 20th century.

Lutz, Peter L. *The Rise of Experimental Biology: An Illustrated History.* Totowa, N.J.: Human Press, 2002. The story of medicine and biology, from their earliest historical origins to the modern day. Lutz starts with prehistoric cave painters, whose work reveals a close observation of animals and the natural world, then follows with accounts of doctors and natural philosophers from antiquity through modern times. The texts are accompanied by beautiful illustrations and photographs.

Magner, Lois N. *A History of the Life Sciences.* New York: M. Dekker, 1979. A detailed history of the evolution of ideas in biology, from ancient times to the modern age, written for nonspecialists.

McElheny, Victor K. *Watson and DNA: Making a Scientific Revolution.* Cambridge, Mass.: Perseus, 2003. A retrospective on the work and life of this extraordinary scientific personality.

Pinto-Correia, Clara. *The Ovary of Eve: Egg and Sperm and Preformation.* Chicago: University of Chicago Press, 1997. A wonderfully readable journey to the worlds of the 17th and 18th centuries and how scientists, theologians, and other scholars of the time attempted to explain conception and very early embryonic development.

Purves, William K., et al. *Life: The Science of Biology.* Kenndallville, Ind.: Sinauer Associates/W. H. Freeman, 2003. A comprehensive overview of themes from the life sciences.

Reeve, Eric, ed. *Encyclopedia of Genetics.* London: Fitzroy Dearborn Publishers, 2001. A collection of articles by experts from different specialized fields in genetics, including excellent discussions of Hox genes.

Sacks, Oliver. *The Island of the Colour-Blind and Cycad Island.* London: Picador, 1996. The famous neurobiologist's story of his trip to Pingelap Island in the Pacific, where a large proportion of the population suffers from a genetic condition that allows them to see only sharp contrasts and shades of gray.

Slack, J. M. W. *Essential Developmental Biology.* 2d ed. Oxford: Blackwell Science, 2006. An excellent textbook covering the main themes of animal development, intended for second- to fourth-year undergraduate university courses.

Stent, Gunther. *Molecular Genetics: An Introductory Narrative.* San Francisco: W. H. Freeman, 1971. A personal account of the early days of genetics by a student of Thomas Hunt Morgan.

Tanford, Charles, and Jacqueline Reynolds. *Nature's Robots: A History of Proteins.* New York: Oxford University Press, 2001. A popular account of the early days of biochemistry and the study of protein structures and functions.

Tudge, Colin. *In Mendel's Footnotes.* London: Vintage, 2002. An excellent review of ideas and discoveries in genetics from Gregor Mendel's day to the 21st century.

———. *The Variety of Life: A Survey and a Celebration of All the Creatures That Have Ever Lived.* New York: Oxford University Press, 2000. A beautifully illustrated "tree of life" classifying and describing the spectrum of life on Earth.

Twyman, R. M. *Developmental Biology.* Oxford: BIOS Scientific Publishers, 2001. A very good overview of the field and useful as a quick reference at the beginning college level, with clear line drawings of the important developmental stages of the main animal model systems. The closing chapter focuses on plant development and contrasts it with that of animals.

Wallace, Alfred Russel. *Natural Selection and Tropical Nature: Essays on Descriptive and Theoretical Biology.* Boston: Adamant Media, 2005. A reprint of a collection of Wallace's major papers and writings, originally published in 1891.

Watson, James D. *The Double Helix.* New York: Atheneum, 1968. Watson's personal account of the discovery of the structure of DNA.

Watson, James D., and Francis Crick. "A Structure for Deoxyribose Nucleic Acid." *Nature* 171 (1953): 737–738. The original paper by Watson and Crick providing the first accurate structure of DNA.

Web Sites and Web Pages

American Society of Naturalists. "Evolution, Science, and Society: Evolutionary Biology and the National Research Agenda." Available online. URL: http://www.rci.rutgers.edu/~ecolevol/fulldoc.html. Accessed December 20, 2008. A document from the American Society of Naturalists and several other organizations summarizing evolutionary theory and showing how it has contributed to other fields, including health, agriculture, and the environmental sciences.

California Institute of Technology. "The Caltech Institute Archives." Available online. URL: http://archives.caltech.edu/index.cfm. Accessed December 20, 2008. Hosts materials tracing the history of one of the United States's most important scientific institutes since 1891. One highlight is a huge collection of oral histories with first-hand accounts of some of the leading figures who have been at Caltech, including George Beadle, Max Delbrück, and many others.

Center for Genetics and Society. "CGS: Detailed Survey Results." Available online. URL: http://www.geneticsandsociety.org/article.php?id=404. Accessed December 20, 2008. This article presents the results of numerous surveys conducted in the United States and elsewhere on topics related to human cloning and stem cell research.

Cold Spring Harbor Laboratory. "Dolan DNA Learning Center's Gene Almanac." Available online. URL: http://www.dnalc.org/home.html. Accessed December 20, 2008. A large collection of multimedia materials for students and teachers on themes such as genes and health, evolution, eugenics, cancer, and the links between genes and cognition. The site hosts hundreds of interviews with prominent scientists who give—at a level that teachers and students can easily understand—an overview of the main questions of biology as well as insights into specific biological processes. Visitors can also conduct virtual experiments and keep track of the latest developments in brain research.

Endowment for Human Development. "Interactive Prenatal Timeline Tutorial." Available online. URL: https://www.ehd.org/timeline_tutorial.php. Accessed December 20, 2008. A rich collection of images, films, and educational materials following embryonic development from conception to birth, with an emphasis on health education for young people and couples planning a family. The site extensively explores the relationship between a mother's health and nutrition and its lifelong effects on her child.

European Molecular Biology Laboratory. "The Digital Embryo." Available online. URL: http://www.embl-heidelberg.

de/digitalembryo/. Accessed December 20, 2008. An amazing series of movies from the laboratory of Jochen Wittbrodt, covering the early embryonic development of the zebrafish. The movies were taken with a three-dimensional microscope setup that could track the division and migration of every cell from fertilization through gastrulation. Through digital reconstructions the scientists are able to track the fate of each cell from its birth to its specialization into organs and tissues.

National Institute of Child Health and Human Development. "The Multi-Dimensional Human Embryo." Available online. URL: http://embryo.soad.umich.edu/. Accessed December 20, 2008. A collection of images of human embryos made through magnetic resonance imaging, covering the major stages of development. Viewing tools allow visitors to scan the embryo "slice by slice" or as three-dimensional animations.

Nature. "Spemann's organizer and self-regulation in amphibian embryos: *Nature* Reviews Molecular Cell Biology." Available online. URL: http://www.nature.com/nrm/journal/v7/n4/suppinfo/nrm1855.html#nrm1855-sx. Accessed December 20, 2008. This Web page presents a reenactment of the famous "Spemann-Mangold experiment," involving transplantation of tissue from one fish embryo into another, leading to an organism with two heads. The experiment is carried out by noted developmental biologist Eddy de Robertis.

Nobel Foundation. "Video Interviews." Available online. URL: http://nobelprize.org/nobel_prizes/medicine/video_interviews.html. Accessed December 20, 2008. A collection of short video interviews with many of the Nobel laureates in physiology or medicine from the last few decades, including many of the leading figures in modern developmental biology.

Sinauer Associates. "Developmental Biology 8e Online." Available online. URL: http://8e.devbio.com/. Accessed December 20, 2008. A vast amount of material related to Scott Gilbert's textbook on developmental biology is available here. As the site says, it is not a textbook, "it is more like a museum. . . .

The material here is loosely based on the theme: 'This is really interesting; it's too bad I can't put it into the textbook.'" Included are stories from the history of developmental biology, details of experiments, and ethical themes.

University of North Carolina at Chapel Hill. "*C. elegans* Movies." Available online. URL: http://www.bio.unc.edu/faculty/goldstein/lab/movies.html. Accessed December 20, 2008. An excellent site from the laboratory of Bob Goldstein at the University of North Carolina. The site and the lab's work are devoted to the development of the worm *C. elegans,* which has been a very important model organism for developmental biology. The site hosts movies of normal development of the worm and its cells; it also shows how these processes are disrupted when scientists interfere with important developmental genes.

Visible Embryo Web site. "The Visible Embryo Homepage." Available online. URL: http://www.visembryo.com/baby/. Accessed December 20, 2008. Visitors can follow the entire history of a human from conception to birth, through all the major stages of development, with illustrations of the embryo and a brief description of what happens at each step.

Index

Note: *Italic* page numbers indicate illustrations.

C

D

U

V

W

X

Y

Z